Physiology Question-Based Learning

Hwee Ming Cheng

Physiology Question-Based Learning

Cardio, Respiratory and Renal Systems

Springer

Hwee Ming Cheng
Faculty of Medicine
University of Malaya
Kuala Lumpur
Malaysia

ISBN 978-3-319-12789-7 ISBN 978-3-319-12790-3 (eBook)
DOI 10.1007/978-3-319-12790-3

Library of Congress Control Number: 2014960134

Springer Cham Heidelberg New York Dordrecht London

Printed on acid-free paper

Springer is part of Springer Science+Business Media (www.springer.com)

Preface

"...Teacher, you have spoken well..they no longer dared to ask him any question"
Luke the Physician (20:39,40)

This book is a first fruit publication of more than a decade of organizing and hosting in Kuala Lumpur, Malaysia the Inter-Med School Physiology Quiz (IMSPQ). This is now a mega physiology event and at the recent 12th IMSPQ, 2014, we gathered 88 medical school teams from 23 countries who came converge for a 2 day adrenaline-high, physiologically stimulating activities.

Physiology questions asked in the competition is the focus of the IMSPQ. Above the friendly tussle for the Challenge Trophy (named in honor of Prof A Raman, the first Malaysian professor of physiology at the University of Malaya), the IMSPQ event is a nucleus for learning and enjoying physiology. The IMSPQ is an invaluable test experience where students of physiology from diverse curriculums of numerous countries are evaluated in the same sitting.

Valuable insights have been gained from a study of the common incorrect responses to the physiology questions asked during both the silent, written and the oral quiz session before a live audience. This book distills some of the major physiological concepts and principles that are part of the IMSPQ challenge. Three systems, cardiovascular, respiratory, and renal are covered, including integrated topics that synthesize essential homeostatic mechanisms of interorgan physiology.

This book is not purposed merely for preparations for teams gearing up for an IMSPQ event. The questions and explanations given, will be a resource for understanding physiology as they highlight the framework and major pillars of physiological knowledge in each system. These questions will provide a good foundation for students to build upon as they continue to pursue the wonders of human physiology.

My appreciation to Thijs van Vlijmen, who from our first meeting, recognized the usefulness of harvesting the IMSPQ for a fruitful book and was enthusiastic in producing this Physiology Question-Based Learning (Pq-BL) series. My student Adlina Athilah Abdullah drew the beautiful flower-blooming heart, lungs, and kidneys (and other illustrations in the text) that introduce the three branches of this PqBL.

At the 12th iMSPQ, we had more than a hundred physiology educators that accompanied their student teams. I hope this book will also be a good teaching tool for lecturers in all their educational efforts to communicate physiology well.

Dr Cheng Hwee Ming Department of Physiology, Faculty of Medicine, University of Malaya, Kuala Lumpur, Malaysia
chenghm@ummc.edu.my

Physiological Flows

I use the hot iron as a painting tool. Movements manipulated by the iron (on which wax paints are applied) are like brushstrokes, for example, shifting and lifting the iron creates wave-like or capillary-like forms. To me, a single movement of the iron signifies a moment in time. It is that single moment, the 'here and now' that holds all reality. With this way of thinking, making an artwork is a very direct, focused, yet intuitive activity. Chew Lean Im

This creative piece by my college friend, Lean Im reminds me of the importance of flow in physiology, including blood flow, airflow, and urine flow. Cheng Hwee Ming

The Questioner and the Questioned (Not the Alligator Interrogator and the Chicken!)

There is much value in using carefully designed questions in teaching. Learning physiology can be improved by the use of well-constructed questions. There are three situational types of question dynamics we can consider: the teacher himself, the teacher–student relationship, and the student learning community.

Teacher as questioner, to himself: self-conversation

a. Why does she misunderstand this mechanism?
b. How can I make her think through this mechanism physiologically?
c. What are the main concepts to convey to my students?
d. What foundational knowledge does she need before she can proceed to understand this mechanism? (*Physiolego* knowledge blocks)
e. How can I reduce mere "swallowing of information" and promote more chewing and thinking through the physiology?

Teacher as questioner to student (Homeostatic teaching)

1. To uncover misperceptions
2. To highlight inaccurate thinking process
3. To stimulate curiosity
4. To strengthen the conceptual learning
5. To guide into integrative thinking on whole body physiology
6. To entrain the ability to apply physiology to pathophysiology

Student to student, peer teaching and "self-directed" learning

The teacher by his planned questioning, model for his students how to think through and enjoy learning physiology among, and by themselves.

Contents

Part I
Cardiovascular Physiology

Introduction: Cardiovascular Physiology

Heart must pump. Blood must flow. These two cardiovascular slogans are the reasons we continue to stay alive. The heart is a rhythmic pump supplying blood in a closed system of flexible vascular conduits. The muscle of the heart (cardiac muscle) is one of the three specialized muscle types in the body besides skeletal and smooth muscles, the latter found in blood vessel wall. The cardiac rhythms are the music of life! To appreciate cardiovascular physiology, a student needs to understand several unique properties of the cardiac muscle pump, including:

1. How action potential is spontaneously generated and transmitted in the heart.
2. The ionic basis of an action potential in the cardiac ventricle muscle.
3. The relationship between electrical activity and mechanical contraction during a cardiac cycle.
4. The role of cardiac autonomic nerves (sympathetic and parasympathetic) on the heart.
5. Factors affecting cardiac output (heart rate × stroke volume) in particular, the separate mechanisms of sympathetic nerve action, and Starling's intrinsic myocardial mechanism.

The circulatory system is functionally two circulations arranged in series. The textbook figures sometimes give the impression that the systemic and the pulmonary circulations are two parallel circuits. In reality, a fixed volume of blood is continuously pumped around in a closed system. The heart can then be seen as two rhythmic pumps (right and left ventricles) contracting synchronously. It is a two-piston engine, ejecting simultaneously two cardiac outputs to the lungs and to the rest of the organs in the periphery. Since the blood volume is a fixed entity, redistribution of cardiac output in response to changing metabolic demands from different organs is part of the homeostatic mechanisms in cardiovascular physiology. Some of the key concepts that a student should focus on include:

1. The role of elastic recoil of the arteries in providing the diastolic blood pressure.
2. Cardiac output and peripheral resistance and determinant of arterial blood pressure.
3. Baroreflex and selective sympathetic vasoconstrictor action on nonessential organs, sparing the coronary and cerebral circulations.
4. The venous capacitance function and role of venous return in cardiac output regulation.
5. Increased cardiac output response during physical activity that involves sustaining a higher blood pressure concurrent with vasodilation of skeletal blood vessels.
6. Special features of blood flow to the rhythmically pumping heart and also to the brain.
7. Role of renal functions and renal sympathetic nerve in blood volume and blood pressure regulation.

Fetal circulation in utero is a special case during our watery beginnings. However, the basic hemodynamics can explain the direction of blood flow in the fetus as well as the conversion from fetal to adult circulation after birth.

Chapter 1
Ins and Outs of the Cardiac Chambers

The flow of blood through the normal, healthy heart is always unidirectional, in both the right and left sides of the heart. This is endured by the sequential, opening and closing of the atrioventricular valves and the aortic/pulmonary valves. Blood flows when there is a pressure gradient. The phasic changes in atrial and ventricular pressure during a cardiac cycle determine, in concert with gating valves, the unidirectional intracardiac flow. The student should understand what generates the pressure that ejects blood volume from each ventricle and what pressure gradient drives the inflow or ventricular infilling of blood during diastolic relaxation phase of the cardiac cycle. The questions below address the physiology of some of these cardiac events.

1. What cardiac index is used as a quantitative measure of myocardial contraction strength?

Answer Myocardial contractility is the term for the power of cardiac muscle contraction and is represented by the ejection fraction.

Concept Cardiac muscle contract as for skeletal muscles. Both muscle types perform work, the skeletal muscles in isotonic contraction and the heart does cardiac work in ensuring a continuous blood perfusion to all the peripheral tissues. The strength of cardiac muscle contraction can also be increased. In the skeletal muscles, graded muscle tension is increased by recruitment of more motor units and higher frequency of motor nerve impulses to produce summative, titanic contraction.

In cardiac muscles, the strength is increased by cardiac sympathetic nerve and circulating hormones, the main one being adrenaline, that binds to beta adrenergic receptors on the cardiac muscles, that are also activated by neurotransmitter noradrenaline released from the sympathetic fibers.

The increased contractility is also represented by the increased ejection fraction. The ejection fraction is the ratio of the ejected stroke volume and the end-diastolic volume (EDV) in the ventricle before contraction. For a given EDV, more volume is pumped out by the more contractile ventricle. The volume remaining in the

H. M. Cheng, *Physiology Question-Based Learning*, DOI 10.1007/978-3-319-12790-3_1

ventricle after systolic contraction, the end-systolic volume will be reduced when myocardial contractility is increased.

In hyperthyroidism, the contractility is also increased by the excess circulating thyroid hormones. Thyroid hormones upregulate beta adrenergic receptors on the cardiac muscle and potentiates the sympathetic/adrenaline positive inotropic actions. Positive inotropism means the same as increased myocardial contractility/higher ejection fraction. Thyroid hormones can also alter the myosin ATPase type in the cardiac muscle, which also accounts for the greater contractility.

2. In Starling's mechanism of the heart, what are the *y*- and *x*-axes of the Starling's graph?

Answer The *x*-axis is the EDV and the *y*-axis is the stroke volume.

Concept The mechanical property of the cardiac ventricle muscles described by Starling is an intrinsic muscle phenomenon. By "intrinsic" this means that no extrinsic nerve or circulating hormones play a role in the Starling's mechanism (or Law) of the heart.

The heart is a generous organ. If it receives more blood volume, it will give out more blood volume. The heart does not hoard! Using cardiac volumes to describe Starling's event, this states that the greater the physiologic increase in the EDV, the bigger will be the stroke volume.

The axes of the graph can also represent *x*-axis as ventricular filling (venous return) and *y*-axis as cardiac output. A larger EDV stretches the ventricular muscle and the contracting tension is greater. The histophysiologic basis for this is the degree of potential overlap between the actin and myosin filaments in the cardiac muscle at different lengths.

Up to an optimal length, the increase in EDV and hence cardiac muscle length will be followed by a larger ejected, stroke volume.

Note that the ejection fraction is unchanged (this fraction is a measure of myocardial contractility, question 1 above).

This "more in more out" ventricular Starling's mechanics applies in both the left and right ventricles. The maximum systolic intraventricular pressure is much higher in the left ventricle (120 mmHg) compared to that in the right ventricle (~30 mmHg). However, the stroke volume (SV) of each ventricle is the same, because the cardiac work has to be higher for the left ventricle against a higher "afterload" (~100 mmHg) than the afterload at the right side represented by the mean pulmonary arterial pressure.

The intraventricular and aortic/pulmonary vascular pressures are different on each side of the heart, but the volume dynamics (EDV and SV) of the Starling's cardiac mechanism is the same and operative in both ventricles.

3. What mechanism ensures that the right and left ventricular cardiac outputs are equalized over time?

Answer Starling's mechanism of the heart has the essential physiologic role in equalization the cardiac output of the two ventricular pumps that are arranged in series.

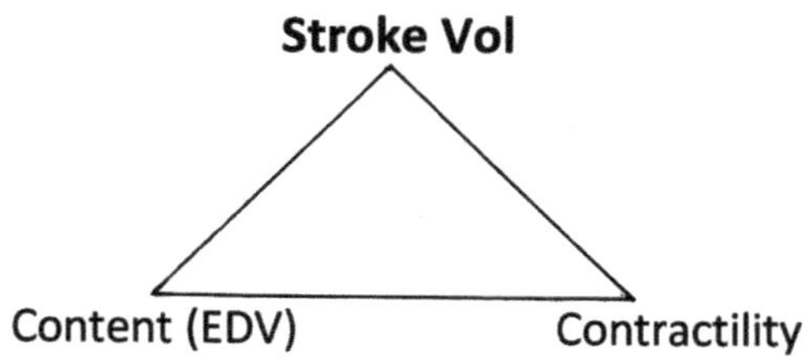

The ejected blood volume with each heartbeat (stroke volume, SV) is determined by two contributing factors. One is an intrinsic cardiac muscle mechanism (Starling's law) where SV is dependent on ventricular filling (*EDV*). The other way to increase SV is by an increased cardiac sympathetic nerve action or by higher circulating adrenaline that both produce a greater myocardial contractility. Increased contractility is defined by a bigger ejection fraction (SV/EDV)

Concept The figure in physiology texts sometime gives the students the impressions that the systemic and the pulmonary circulations are in parallel. If we imagine stretching out the whole circulatory system into a chain, the right and the left ventricles would be seen to be connected in series like two beads along the "bloody" vascular chain.

The serial arrangement of the right and left ventricular pumps present a potential problem for the vascular blood traffic flow. It is crucial that the two pumps are synchronized with regards to the cardiac output "put out" be each ventricle. Should there be unequal cardiac outputs, very soon we will have problems of vascular traffic congestion.

Beat by beat, there could be small fluctuations in the stroke volume from each ventricle. There are 60 beats in each minute, and the differences in the stroke volumes will accumulate to produce unequal cardiac output if there is no mechanism to adjust for this.

This is where the intrinsic Starling's myocardial mechanics becomes important. If one ventricle has a larger cardiac output, this will mean a greater filling of the other ventricle. The second ventricle then contracts more strongly. If the second ventricle does not intrinsically pumps out more of what it has received (increased EDV), then the traffic "upstream" from the second ventricle will be congested.

To give a clinical illustration, if the right ventricle weakens, there will be venous congestion (with development of peripheral edema). On the other "hand" ("heart"), left ventricular failure will result in pulmonary venous congestion and this can cause pulmonary edema.

"O my Starling, my heart(s) beat for you!"

4. In a transplanted heart, how may cardiac output be increased during physical activity?

Answer The cardiac function of the denervated transplanted heart responds to circulating hormones.

Concept The life-giving heart can be donated. The heart has autorhythmicity, the sinoatrial (SA) pacemaker cells spontaneously generating action potentials that are transmitted throughout the myocardium.

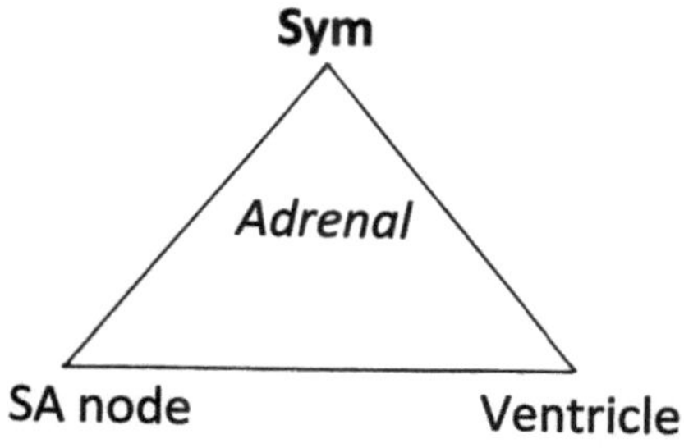

The sympatho-adrenal medullary axis supplements the direct sympathetic effects on the heart. The adrenergic receptors at the sinoatrial node and the ventricular muscles bind to circulating adrenaline besides binding the neurotransmitter noradrenaline released from cardiac sympathetic nerve terminals

The normal heart does not require extrinsic neural innervation to maintain its cyclical beats or contractions. The pacemaker activity is increased by sympathetic input that produces the tachycardia during exercise. In the transplanted heart, the SA node can still be stimulated by circulating adrenaline from the adrenal medulla.

Cardiac output is the product of the heart rate and the stroke volume. The ventricle contraction of the heart can also be strengthened by adrenaline. Adrenaline increases both the heart rate and the contractility (increased ejection fraction) of the heart.

In the overall circulation, it is natural to view the heart as the center of all functions. This cardiocentric concept of blood circulation physiology may hide the important key contribution of venous return in the closed circuit of the cardiovascular system. The heart only pumps out what blood volume fills it, and the operating blood volume is a fixed entity.

Thus, the venous return is certainly an important provider for the increased cardiac output during physical activity in a person with a donor's heart. Venous return is increased during exercise by several factors including muscle pump effect and respiratory pump effect of central venous pressure. Sympathetic venoconstriction also decreases the venous capacitance, so more blood is available to circulate.

5. What ensures that the atrial and ventricular contractions are orderly and sequential?

Answer The slight transmission delay at the atrioventricular node allows the ventricular systole to proceed only after the artila systole.

Concept The left and right ventricles contract simultaneously. The ventricles function together like a syncytium. The right and left atria also contract as a functional syncytium. The cardiomyocytes in both the ventricles and the atria are electrically coupled via gap junctions, besides being spread of the action potentials by the conducting fibers.

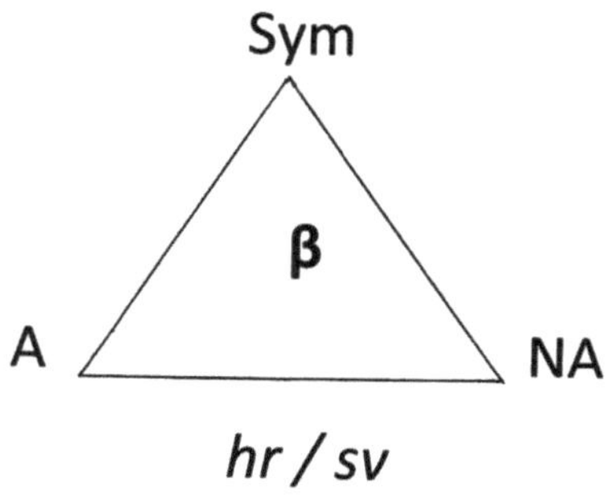

The sympathetic nerve activates the beta receptors (beta looks like a standing heart!) on the sino-atrial node and the ventricular muscles to produce tachycardia and increased myocardial contractility that ejects a larger stroke volume. The sympathetic neurotransmitter is noradrenaline. Secretion of the adrenal medullary catecholamine, adrenaline is also stimulated by sympathetic cholinergic nerve. Adrenaline binds and acts on the cardiac beta receptors

Atrial systole occurs during the final stage of ventricular diastole when the ventricles are filled with blood. The EDV is achieved by both passive ventricular infilling of blood and a "top-up" by atrial contraction.

If is thus imperative that the ventricles are not depolarized too soon after atrial depolarization. This will allow the atrial systole to fill the ventricles before the ventricles contract.

The transmission of impulses from the sino atrial pacemaker through the atrial muscle is slightly slowed at the atrioventricular node, the only transit electrical point between the atria and the ventricles. The atrioventricular "delay" ensures that atria depolarization and generation of action potentials are near completed before the ventricles become depolarized (the "P" wave is temporally separated from the "QRS" complex).

6. How does the function of the cardiac/vascular valves signal the different phases of the cardiac cycle?

Answer Closure of atrioventricular valve begins the systolic phase of the cardiac cycle and closure of the pulmonary/aortic valves signal the start of diastole.

Concept The cardiac cycle of the rhythmic beating heart is divided into the ventricular filling phase during diastolic relaxation and ventricular systolic contraction phase. The cardiac valves ensure that the intracardiac flow of blood is unidirectional, only from the atria into the ventricles.

When the ventricular muscles are depolarized, the mechanical contraction develops. As the ventricular muscle tension starts to increase, very soon the intraventricular pressure exceeds the atrial pressure. The mitral and tricuspid valves at the left and right side of the heart, respectively, snap shut. This produces the first heart sound. This begins the systole and the initial brief period of systole is an isovolumetric contraction when the intraventricular pressures build up steeply until the point when the pulmonary/aortic valves are forced open during the ejection phase.

When the ventricles are repolarized, this will relax the muscles. When the intraventricular pressures drop to less than the pulmonary/aortic arterial pressures, the pulmonary/aortic valves shut. This produces the second heart sound. Backflow of the pulmonary and aortic blood in to the ventricles is prevented.

The closure of the these valves begins the diastole, and the initial period of diastole is the isovolumetric relaxation when the intraventricular pressure drops precipitously until the tricuspid and mitral valves open for ventricular filling.

When the cardiac or vascular valves do not close completely, this is termed a valvular insufficiency. An insufficiently shut valve will result in a heart murmur. For example, if the left mitral valve is insufficient, contraction of the left ventricle will squirt blood flow abnormally back into the atria. A systolic murmur is heard during the first heart sound.

If the aortic valve is insufficient, the back flow of blood into the left ventricles during diastole occurs. A diastolic murmur is heard at the second heart sound.

On the normal electrocardiogram (ECG), it would benefit the student to attempt to reason and derive that the first heart sound is located just after the QRS ventricular depolarization wave, and the second heart sound is placed just after the T-ventricular repolarization deflection.

7. What are the two pressures that determine the ventricular filling of blood from the systemic circulation?

Answer Venous return is driven by the perfusion pressure which is the difference of the mean circulatory (systemic) filling pressure and the right atrial pressure (rap).

Concept The rap is functionally synonymous with the central venous pressure. The mean systemic filling of circulatory pressure (msfp) is the average pressure in the systemic circulation that determines the venous return. Experimentally, the msfp is obtained by acutely stopping the heart of the animal from beating. The rap is then measured. Since the cardiac output is now zero, the average pressure in the systemic circulation and the rap must be the same. This is what is conceptually called the msfp.

From the basic hemodynamics equation, the flow is equal to the perfusion pressure/vascular resistance. When we consider venous return, this will translate to

Venous return = msfp minus rap/venous return.

Since venous resistance is small in contrast with arterial resistance, venous return is basically conditioned by the msfp and rap.

To illustrate with clinical situations, right ventricular failure will raise the central venous pressure. Venous return is impeded and venovascular congestion develops. Hypovolemia from any causes decreases the msfp resulting in reduction of venous return and cardiac output.

Doing a Valsalva maneuver (e.g., include straining at stools, exertion during labor) increases the intrathoraic and central venous pressure. The perfusion pressure to deliver venous return becomes smaller. A similar situation of increased central venous pressure would be in patients maintained on positive pressure breathing.

This Chinese pictogram of "heart" resembles the cardiac ventricles. The *extreme left stroke* would then represent physiologically the venous return and the *far right stroke* the cardiac output. In a closed circulatory system, the venous return would equal the pulmonary blood flow (right cardiac output) and the rate of ejected blood flow from the left ventricle into the systemic circulation

During exercise, the venous return is enhanced, since deeper tidal volume breathing decreases the rap. This is described as a "respiratory pump" effect.

8. Is there a proportionate relationship between heart rate and cardiac output?

Answer At high tachycardia, the decreased diastolic filling time tends to reduce stroke volume, and so the cardiac output does not increase linearly with increase in frequency of heart beat.

Concept The heart pumps out only what it contains. The volume of blood pumped out per beat (stroke volume) is determined by both the EDV and the myocardial contractility (increased by sympathetic nerve/adrenaline).

The diastolic period is more significantly reduced than systole during tachycardia when the cardiac cycle is shortened. This has the effect of reducing the EDV.

We can then expect that since the cardiac output is the product of heart rate and stroke volume, the cardiac output will not increase proportionately with increasing frequency of heart beats.

The student should not mix up the effect of heart rate on stroke volume and the cardiac output. The stroke volume could be lessened due to the reduced ventricular filling and thus the EDV. However, the cardiac output is still more than the value at rest.

The student should be reminded that whenever tachycardia occurs, the cardiac sympathetic nerve is stimulated (concurrent with a decreased vagal parasympathetic activity to the pacemaker cells).

This means that the sympathetic tachycardia as in exercise is always concurrent with a positive inotropic effect of sympathetic action on the ventricular contractility (the sympathetic nerve releases adrenaline also from the adrenal medulla). What this indicates is that although the EDV is less, the ejection fraction is enhanced by the increased myocardial contractility. The net effect is that the stroke volume may not be that much decreased during greater cardiac activity. In a situation when only the tachycardic reduction of the EDV is considered, the effect on the cardiac output will theoretically be more, if sympathetic action on producing a higher ejection is ignored.

9. How does venous return directly influence the heart rate? Bainbridge reflex

Answer Increased venous return produces an increased heart rate to help maintain an optimal rap.

Concept The rap or central venous pressure (cvp) is near 0 mmHg. This rap (cvp) fluctuates during a normal respiratory cycle, slightly lower during inspiration compared to expiration. The venous return graph has the rap as the *x*-axis and venous return as the *y*-axis. The graph shows an inverse relationship between the rap and venous infilling of the heart.

In the venous return graph, rap is the cause and venous return flow is the affected factor. Since the circulation is a closed system, it is also true that venous return changes as a causative factor effect and alter the rap. This venous return/rap coupling explain the reflex response to increased venous return on producing a tachycardia. This is also named the Bainbridge reflex.

The student who is familiar with the baroreflex will wonder at the integration between the Baindridge and the carotid/aortic baroreflex. Any increase in venous return would lead sequentially to an increased cardiac output and arterial blood pressure. The typical baroreflex to the increased blood pressure is a bardycardic response.

Thus, we have a direct tachycardic effect of increased venous return and an indirect bradycardic effect of a higher venous return via the baroreflex mechanism.

The text in Boron's *Medical Physiology* proposed that the Bainbridge effect is more prominent in hypervolemia in order to prevent an elevation of rap or central venous pressure. This could potentially cause venous vascular congestion. In hypovolemia/hypotension, the compensatory, baroreflex/sympathetic effector action is given priority.

10. How is the hemodynamics of a rhythmic pump different in terms of flow and pressure?

Answer For the rhythmic cardiac pump, the flow or cardiac output is not proportionate to the pressure as the aortic blood pressure is also the "afterload" against which the rhythmic pump contacts.

Concept For vascular flow, the basic hemodynamics apply where flow = perfusion pressure/vascular resistance of that segment of the circulation. When the blood flow of the overall systemic circulation is considered, the rhythmic nature of the cardiac pump changes the hemodynamics.

We could still consider the left to right heart flow (cardiac output) as the pressure difference divided by the total peripheral resistance. The pressure difference or driving pressure would be the difference between the aortic and the rap. We cannot use the left intraventricular pressure in the hemodynamics of left/right heart flow as the cardiac pump is rhythmic, and intraventricular pressure during diastole is close to 0 mmHg.

The aortic or mean arterial pressure is the "head" pressure for providing the continuous flow to the periphery. When the ventricle contracts and pumps from relaxed position, the ventricle has to pressurize against the aortic blood pressure to produce blood flow. The aortic pressure represents the "afterload" (the afterload of the right ventricle is the pulmonary arterial pressure).

In hypertension, the "left" afterload is elevated, and more cardiac work has to be done to pump to perfuse the peripheral tissues. The chronic overload on the left ventricle leads to ventricular hypertrophy. Likewise, in pulmonary hypertension, the right ventricle is burdened with extracardiac work. Right ventricular hypertrophy develops.

Chapter 2
Cardiac Cycle

The rhythmic heart repeatedly pumps, relaxes to "top-up" and pumps. In a cardiac cycle, there are two main phases, the contraction (systole) and the relaxation of the ventricles (diastole). Although the atria also have their own cycle of similar contractile activity, the use of the words systole and diastole refer to the ventricles that eject blood out with each stroke volume. There are cyclical intraventricular pressure and volume changes. The pressure/volume changes can be matched to the electrical activity that starts at the sinoatrial pacemaker cells and its sequential transmission and spread across the whole myocardium (electrocardiogram, ECG). In addition, the profile of pressure changes in the atria, ventricles, and aorta/pulmonary artery which is associated with opening and closing of valves, the latter generating the major first and second heart sounds. The changes in aortic blood pressure during a cardiac cycle represent the peak systolic blood pressure and the minimum diastolic blood pressure. Understanding the cyclical changes in these parameters takes time, to ponder the step by step cardiac events.

1. Why is the P wave of a normal ECG always smaller than the QRS complex?

Answer The amplitude of the deflections of a normal ECG is determined by the mass of the tissue that has been depolarized/repolarized.

Concept The ECG is a measurement of the electrical activity on the surface of the body. The ECG tracing is not the same as action potential electrical changes of the membrane potentials. The ECG recorded does result from the spread of action potentials through the heart.

The heart is in a conducting medium and electrical currents generated around the surface of the heart as it is being progressively depolarized are transmitted to the body's surface.

If we look at the scale of an action potential, the amplitude is ~100 mV. The amplitude of the major ECG wave, the QRS complex is less than 2 mV.

The mass of cardiac muscle that is "electrified" by the spreading action potentials will determine the size of the electrical currents generated. Therefore, the atrial electrical activity during a cardiac cycle will produce a smaller deflection than the larger ventricles.

H. M. Cheng, *Physiology Question-Based Learning*, DOI 10.1007/978-3-319-12790-3_2

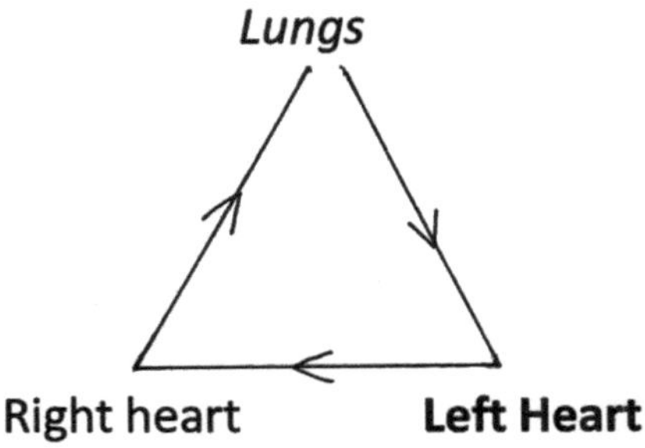

The clockwise arrow direction indicates the unidirectional blood circulation through the left ventricle and the right ventricle, both ventricles are in series, with the lungs in between. The rate blood flow from the more muscular left ventricle (cardiac output) must be equalized with the right ventricular cardiac output to avoid any vascular "bloody traffic" congestion

Note that the smaller amplitude of the ECG "P" wave is not that the atria contract less strongly than the ventricles. It is also not explained by the smaller volume size of the atria.

When there is an increase in the mass of a cardiac chamber, this is then reflected in the ECG deflection. In ventricular hypertrophy, the amplitude of the QRS wave will be bigger.

2. How does the parasympathetic nerve affect the P–R interval and the R–R interval?

Answer Parasympathetic nerve acts to increase the duration of both the R–R and the P–R intervals of the ECG.

Concept The heart rate is spontaneously generated by the pacemaker activity of the sinoatrial (SA) nodal cells. These action potential self-generating cells have dual autonomic control from the parasympathetic and the sympathetic nerves.

The normal resting heart rate is due to a dominant vagal parasympathetic input. If this vagal chronotropic tone is reduced, tachycardia occurs.

The R–R interval is one cardiac cycle, from one ventricular depolarization to the next. A tachycardic effect will decrease the R–R interval.

From the SA node, cardiac impulses are transmitted synchronously through the atrial functional syncytium. The cardiac impulse is slightly "delayed" at the atrioventricular (AV) node to allow for sequential atrial and ventricular contractions.

The AV node is the sole electrical conduction pathway from the atria to the ventricles. In the normal ECG, the P–R interval represents the time taken for the cardiac impulse to be transmitted from the beginning of atria depolarization to the initiation of ventricular depolarization.

Most of the P–R interval is the time transit at the AV node.

The AV node is also innervated by parasympathetic fibers. Parasympathetic impulses to the AV node slow the impulse transmission. The P–R interval is lengthened.

3. Which portion of the normal ECG accounts for the long electrical refractory period of ventricular muscle?

Answer The prolonged depolarization of the ventricle, as thus the longer refractory period, coincides with the ST segment of the ECG.

Concept The cardiac ventricles have a unique electrical profile of their action potential. There is a prolonged depolarization phase (or delayed repolarization). The ventricular action potential has thus a plateau phase when the ventricle cardiomyocites remain depolarized.

This extended action potential also means that the electrical refractory period of the ventricles is also prolonged. This property protects the cardiac muscle pump from a tetanic contraction. A heart that goes into tetanic contraction will not be filled and the essential perfusion to the brain and the heart will be cut off during the abnormal, sustained contraction.

The QRS wave represents the depolarization event of the ventricles and the T-deflection, the ventricular repolarization. Thus, the time period between the de- and the beginning of the T repolarization wave is the prolonged depolarization seen as the plateau phase of the ventricular action potential. This is the ST segment.

By convention a "segment" of an ECG does not include a wave, while an ECG wave is part of an "interval" period.

This ST segment is thus. When the calcium ions from the extracellular fluid influx into the ventricular cardiomyocytes. The additional calcium cation influx is the reason for the delayed repolarizaton of the ventricles. The entry of extracellular fluid (ECF) calcium into the cytoplasm of the ventricular muscle fibers triggers more calcium release from the sarcoplasmic reticulum (SR). This ECF calcium-SR calcium trigger is described as "calcium induced calcium release."

4. How would you expect the increased circulating adrenaline to affect the QRS amplitude?

Answer Adrenaline should not alter the amplitude of the QRS complex.

Concept The amplitude of the ECF waves is dependent on the mass of the cardiac tissue where the electrical action potential event has occurred. The strength of cardiac muscle contraction is not reflected in the ECG electrical profile.

Adrenaline increases both the heart rate and the myocardial contractility. The R–R interval and the P–R interval will be shortened as the catecholamine binds to the same beta receptors that are bound by noradrenaline released from the cardiac sympathetic nerves.

However, the increased stroke volume due to the positive inotropic effect of a greater cardiac ejection fraction cannot be derived from looking at the ECG. A greater strength of contraction produced by adrenaline action does not increase the amplitude of QRS deflection.

Only in case of ventricle hypertrophy and a more cardiac muscle mass does the ECG inform us by a bigger amplitude of the QRS.

Adrenaline also does contribute to the coronary vasodilation when the heart is more active. Increased coronary perfusion during exercise to supply the greater metabolic demands of the cardiac muscle is not registered either by exercise ECG.

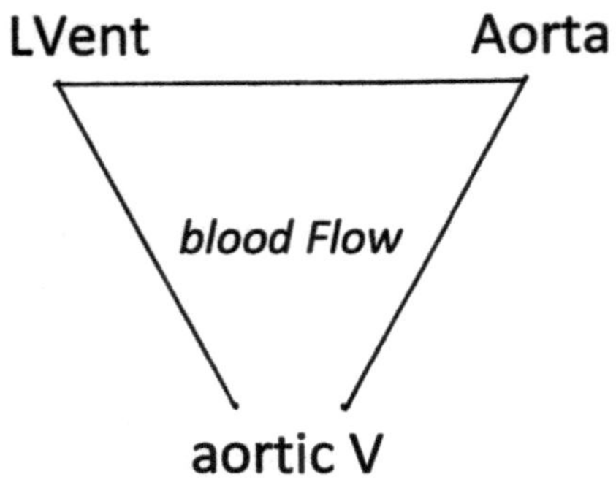

The aortic valve opens and shuts depending on the pressure gradient between the left ventricle/aorta. When left ventricle (*LV*) exceeds the pressure in the aorta, the valve opens, and ejection of a volume of blood (stroke volume) enters the aorta. If the aortic pressure exceeds that in the *LV*, the aortic valve shuts and begins the diastole of a cardiac cycle. The *LV* is filled with oxygenated blood during diastolic filling with pulmonary venous blood from the lungs

However, the converse condition of coronary ischemia can produce some characteristic ECG changes.

5. When does iso-volumetric relaxation occur along the ECG?

Answer Isovolumetric relaxation is the initial brief phase of ventricular diastole and begins just after the T repolarization wave along the ECG.

Concept The electrical event precedes the mechanical event in the heart. The first short stage of diastole occurs when the aortic/pulmonary valves shut. The intraventricular pressure drops markedly during this phase when all the valves including the tricuspid/mitral valves are closed.

Diastolic ventricular filling starts when the tricuspid/mitral atrioventricular valves open.

Ventricular diastolic relaxation occurs after the ventricles are repolarized. Thus, diastole begins just after the T wave. Diastole will proceed until the beginning of systole when the ventricles begin to contract and shut the tricuspid/mitral valves. This point is just after the QRS wave. The period of diastole during a cardiac cycle is then from the end of T wave to the end of the QRS complex.

Thus, the systolic isovolumetric contraction begins just after the QRS deflection and ends when the aortic/pulmonary valves are pressurized open during ejection phase of systole.

6. How does tachycardia affect the myocardial contractility? (Note the opposite effects to no 9.)

Answer Tachycardia has an indirect effect in increasing myocardial contractility via the elevation of intracellular calcium in the ventricular myocytes.

Concept When the heart is more active, it pumps more frequently (tachycardia) and more strongly (increased ejection fraction). Both these chronotropic and inotropic actions are effected by the cardiac sympathetic nerves and adrenaline, respectively.

There is some additional increased cardiac contractility that results from a higher frequency of heart beat. Each cardiac cycle of contraction is followed by relaxation which is initiated when calcium ions are pumped by Ca-ATPase back into the sarcoplasmic reticulum (SR) or extruded by the cell membrane Na/Ca exchanger into the ECF.

This lowering of the cytosolic calcium precedes cardiac muscle relaxation. With tachycardia, there is relatively less time to reduce the intracellular calcium to resting, precontraction level.

During the next muscle depolarization event, there will be a higher intracellular calcium when calcium from ECF influx is released from SR. The myocardial contractility is increased proportionately to the rise in cytosolic calcium.

Note that in skeletal muscles, graded increase in contraction strength is not mediated by increasing intracellular calcium. Skeletal muscle tension is increased by activation of motor unit recruitment and a higher frequency of impulses in the alpha motor neurons that innervate the muscles.

7. How does right heart failure lead to development of peripheral edema?

Answer In right heart failure, the venous congestion leads to increased capillary hydrostatic pressure and higher net capillary filtration that results in fluid accumulation in the interstitium.

Concept In a normal heart, the stroke volume (SV) is proportionately related to end-diastolic volume (EDV, preload) over a physiologic range. In the right side of the heart, this EDV/SV pairing ensures that the central venous pressure/right atrial pressure is consistently at a low value to maintain optimal ventricular filling.

When the right ventricular function is compromised, the ejection volume of the preload is not effectively pumped out. With time, venous congestion develops with elevated central venous pressure. The increased venous pressure will "radiate" and be transmitted into the microcirculation. This backlog of vascular pressure easily occurs because unlike the presence of high resistance precapillary arteriole, the postcapillary venular resistance is low.

At the capillary, the raised capillary hydrostatic pressure will disturb the balance of Starling's forces. The net filtration along the capillary will soon exceed the capacity of the lymphatic drainage to maintain a low interstitial hydrostatic pressure. Fluid accumulates in the tissue spaces (edema).

8. How does the ventricular volume/pressure diagram illustrate the dynamic changes during a cardiac cycle?

Answer The ventricular x-axis volume/y-axis pressure diagram demonstrates the beginning and ending of each of the four phases within the systole and diastole of a cardiac cycle.

Concept The ventricular volume along the x-axis will be the same for the right and left ventricle. However, the y-axis intraventricular pressure will be on a different scale (maximum for left ventricle is 120 mmHg and for right ventricle is ~30 mmHg).

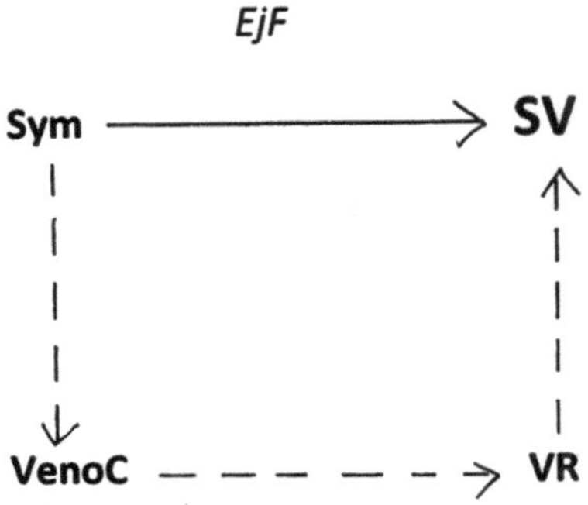

Cardiac sympathetic action increases the stroke volume directly by increasing the ejection fraction. Indirectly, the reduced venous capacitance with sympathetic venoconstriction will also increase venous filling that will produce a bigger stroke volume. *SV* stroke volume. *EjF* ejection fraction. *VR* venous return

Sharp changes in pressures are clearly evident in the two vertical sides of the volume/pressure (V/P) diagram. The diagram must also be followed anticlockwise. This means that the right vertical line shows a steep increase in pressure and the left vertical, a precipitious drop in ventricular pressure. These two pressure lines at constant ventricular volumes would represent the isovolumetric systolic contraction and isovolumetric diastolic relaxation phase.

The maximum volume in the V/P diagram is the end-diastolic volume, EDV (~ 120 ml) and the miminum ventricular volume is the end-systolic volume (ESV, around 40 ml). The lower horizontal line that links the two verticals is thus the stroke volume SV (EDV–ESV). This bottom horizontal line with little change in ventricular pressure, moving anticlockwise from ESV to EDV is the event of ventricular filling.

The south east (SE) corner of the V/P diagram, the start of the vertical rise in pressure is the beginning of systole. The isovolumetric contraction phase begins when the atrioventricular valve shuts and pressure builds up rapidly before the aortic/pulmonary valve is pressurized open. Thus, the SE is where the mitral/tricuspid valves closes.

The upper line that extends from the upper right (NE) to the upper left (NW) corner of the V/P diagram represents a decrease of ventricular volume from EDV to ESV. This is the ejection phase when the ventricles pump out each of their stroke volumes. The NE point is when the pulmonary/aortic valves are opened.

Beginning at the NW corner of the V/P diagram, the intraventricular pressure drops rapidly. The NW corner spot is the beginning of diastole when the aortic/pulmonary valves shut. There is no change in volume, this initial brief phase of diastole until the intraventricular pressure falls to below the atrial pressure. When this is reached, tricuspid/mitral valves open (bottom, left SW corner) and diastolic ventricular filling proceeds.

9. How is the increased myocardial contractility and Starling's mechanism of the heart represented by the ventricular/pressure loop diagram?

Answer Sympathetic nerve action on the heart increases the ejection fraction and the end-systolic volume (SW corner) is shifted to the left. Starling's intrinsic car-

The heart is functionally two rhythmic pumps arranged in series. The contraction phase (systole) and relaxation phase (diastole) are marked by heart sounds due to the closure of cardiac and arterial (pulmonary, aortic) valves. Systole is the period from the first to the second heart sound, and the longer diastolic ventricular relaxation/filling phase runs from the second to the first heart sound

diac muscle contraction mechanism is effected when the ventricular filling or end-diastolic volume (SE corner) is increased or shifted to the right.

Concept For both cardiac sympathetic and Starling's effects, the stroke volume is increased. The higher SV for sympathetically—increased contractility results in less volume (ESV) remaining in the ventricle, although the precontraction EDV is not changed.

For the Starling's effect, the greater stroke volume is due to a bigger volume that fills the ventricle before contraction (higher EDV), stretching the ventricle to a greater systolic muscle tension. The ESV is not altered by Starling's. The increased preload or EDV produce the larger stroke volume.

In the situation when the aortic pressure is elevated (afterload), how would the ventricular volume/pressure loop look like? In chronic hypertension, the left ventricle has to generate additional cardiac work force to pump the same stroke volume against the raised afterload. For a given EDV, if the ventricle begins to weaken, the SV will decrease and the remaining ESV will be more, shifted to the right.

The vertical rise in pressure during the isovolumetric contraction phase of systole (right vertical of V/P loop) will also be greater (contraction against a higher afterload) before the aortic valve opens for the ejection of stroke volume.

10. Why does the venous return curve have the same *x*-axis as the cardiac output curve, drawn on the same graph?

Answer The venous return (*y*-axis) is quite obviously related to the *x*-axis right atrial pressure (rap). Venous return as a primary cause proportionately affects the rap, and thus the expected *x*-axis venous return for the cardiac output (*y*-axis) curve can be harmonized by using the rap as the *x*-axis also.

Concept The circulatory system is a closed circuit. Conceptually, the cause and effect in a particular cardiovascular event can quite often by puzzling and confusing to the students. The cardiac output or cardiac function curve relates the venous return (x-axis) to the cardiac output. This is basically describing the intrinsic Starling's mechanism of the heart, where the ventricular stroke volume proportionately changes with the EDV.

In the presence of cardiac sympathetic activity, the cardiac function curve is shifted to the left; an indication that for the same EDV a positive inotropic effect of sympathetic stimulation gives a bigger stroke volume.

What is the rationale for substituting the x-axis venous return with right atrial pressure?

The students would have heard that changes in the right atrial pressure (central venous pressure), as a cause, would be expected to cause an inverse effect on venous return since the driving pressure for blood entry into the heart would have been reduced (the inverse relationship is seen in the venous return curve).

Because the circulatory system is a closed flow system, it is also true that if we consider venous return as the cause, then the right atrial pressure would change proportionately with the magnitude of the venous return. The Bainbridge effect of increased venous return producing a reflex tachycardia is due to the increased right atrial pressure.

The combined cardiac function and venous return (or vascular function) curves intersect at a certain right atrial pressure. In the closed circulatory system, this equilibrium rap is thus a net value from the dynamics of inflow venous return and outflow cardiac output.

When we consider the cardiac output, functionally, a normal ventricular pumping maintains an optimal rap (we have just discussed above the reverse interactions that the rap affects cardiac output). In right/left ventricular failure, the right/left atrial pressure begins soon to be elevated.

Chapter 3
Blood Pressure

Remember the Giraffe when you think of blood pressure regulation! Gravitational force affects the hydrostatic pressure of blood and the long-neck animal certainly must ensure a constant adequate arterial blood pressure for its cerebral circulation, as it walks around feeding on leaves on high trees.

1. Are the location of the arterial baroreceptors important or can the receptors be located elsewhere e.g. in the abdomen?

Answer The location of the carotid and aortic baroreceptors are not irrelevant to their essential functions. Arterial baroreceptors monitor blood pressure and ensure that blood perfusion especially to the brain is adequate. For humans that spend a considerable of their daily activity in the upright position, the maintenance of a normal optimal blood pressure fulfils the role of protecting the brain from cerebral ischemia.

The location of the carotid/aortic mechanoreceptors above the level of the heart allows the baroreceptors to detect the initial drop in blood pressure when a person stands up from the horizontal resting position. The hydrostatic blood pressure decreases above the heart's level due to gravitational force. If the baroreceptors are located below the heart's level, they will not sense the initial reduction in blood pressure that is a consequence of venous pooling which lowers the stroke volume.

The carotid/aortic baroreceptors are in the high-pressure arterial side of the circulation. They are also called high-pressure volume sensors. This is to distinguish the baroreceptors from volume sensors in the low-pressure, venous side of the systemic vascular circuit. These volume receptors are located in the atrial cardiac chamber and the pulmonary vasculature. They function to monitor the "fullness" of the circulation. The afferent impulses from the volume receptors are, like the impulses from the baroreceptors, sent to the brain stem cardiovascular regulatory neurons.

Afferent impulses from both baroreceptors and volume receptors also input into the hypothalamus to affect antidiuretic hormone (ADH) secretion and thirst sensation. This is part of the water balance control.

In extracellular fluid (ECF) volume expansion, the hypervolemia stretches the volume receptors in the atria and this releases natriuretic hormone to increase sodium and water excretion. Changes in blood volume (determined by total body

H. M. Cheng, *Physiology Question-Based Learning,* DOI 10.1007/978-3-319-12790-3_3

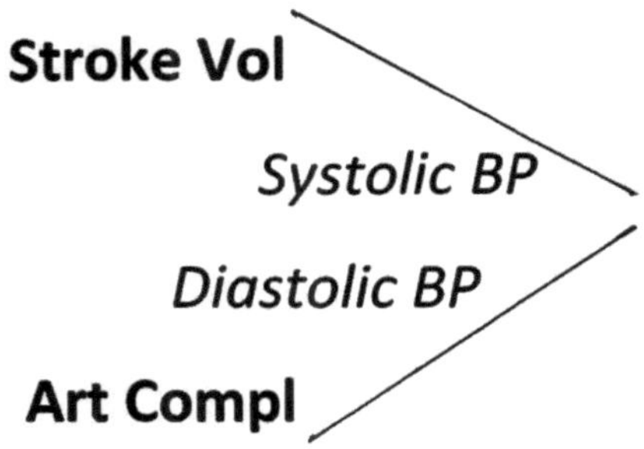

The value of systolic blood pressure (SBP) depends on the stroke volume (SV) and the arterial compliance. A bigger SV gives higher SBP for a given arterial compliance. Similarly for a given SV, a decreased arterial compliance will produce a higher SBP. The diastolic BP is due to arterial recoil. A higher SV will be accompanied by more elastic recoil. A reduced arterial compliance will lower the diastolic blood pressure (DBP)

sodium or sodium balance) will bring corresponding change in the blood pressure via the cardiac output factor.

2. Which determinant of the blood pressure equation does venoconstriction affect?

Answer Venoconstriction by sympathetic nerve increases venous return and this gives a higher cardiac output since a bigger stroke volume will be produced from a larger end-diastolic volume.

Concept Vasoconstriction as a general term means both constriction of the arterial and the venous blood vessels. To understand the differential vascular functions of arteries and veins, it is useful to specify vasoconstriction when thinking about arteriolar increased resistance which primarily determines the total peripheral resistance (TPR).

The term venoconstriction has a different meaning as the venous effect is not so much the contribution to TPR, but the reduction in the venous "blood reservoir". At rest, if you snap a photograph of the whole cardiovascular system, more than 60 % of the total blood volume is in the veins. The veins are supplied by sympathetic nerves and the venoconstriction decreases the venous capacitance. More blood is made available to circulate and fills the heart. A greater cardiac output is pumped by the heart.

Rhythmic muscle contraction during physical activity like walking, jogging also compresses the veins and increases the flow of venous blood via the inferior vena cava into the heart.

3. How is sodium balance and arterial blood pressure physiologically connected?

Answer Sodium balance essentially affects the ECF and blood volume and blood volume is a determinant of arterial blood pressure.

Concept There is quite a lot of popular reports and health advice that too much salt intake is not good and predispose the person to developing hypertension. The association and any cellular mechanisms that link the electrolyte to blood pressure is still being studied and there may be individual salt-sensitivity in hypertensive-prone persons.

Blood pressure is determined by both the cardiac output and the peripheral arterial resistance. There may be vascular responsiveness that is modulated by salt.

Blood volume as part of ECF volume is determined by the sodium balance or the total body sodium. Changes in blood volume affect the mean systemic filling (circulatory) pressure. This value is less than 10 mmHg (students should not confuse this similar sounding term to systemic arterial blood pressure which is ~100 mmHg).

The mean systemic filing pressure is the overall, average driving perfusion pressure for the venous return, since the central venous/right atrial pressure is about 0 mmHg.

Control of blood pressure involving regulation of blood volume is described as "long-term" blood pressure control. Characteristically, blood volume homeostasis involves several anti-natriuretic (e.g., aldosterone) and natriuretic hormones. Actions of hormones require more time compared to rapid, neural reflex actions. "Short-term" blood pressure control is mediated by the baroreflex/brainstem/autonomic sympathetic effector feedback pathways.

4. How are blood pressure, ECF volume, and effective circulatory volume (EfCV) related?

Answer In a normal person, any increase in ECF volume will also increase the EfCV, and this will raise the blood pressure.

Concept The EfCV is a concept and is not a measurable entity or blood parameter. Simply it refers to the blood volume that is effectively perfusing the tissues. Blood must flow for the blood to have any physiologic benefit or meaning to the cells.

Stagnant or sluggish blood flow will insufficiently meet the energy demand of the cells and tissue ischemia results.

In a normal person, the blood volume and the EfCV are functionally the same when the cardiac function is normal. However, if there is cardiac dysfunction, the heart has pump failure and the cardiac output is reduced.

Therefore, although the blood volume in the person with cardiac failure is normal, her EfCV is decreased. Her tissues will experience stagnant hypoxia.

When we consider blood volume sensing, a more specific parameter that is monitored should rightly be the EfCV. To explain, in the above patient, the blood volume is normal (euvolemia), but the volume/pressure receptors detect a drop in the EfCV.

This, for example, is sensed by the intrarenal baroreceptors at the afferent arteriole. This reflexly releases renin from the arteriolar granular juxtaglomerular (JG) cells. The plasma renin then activates sodium-conservating mechanisms and water retention follows. The pathophysiologic outcome is an expanded ECF and blood volume.

In this isotonic expansion in the cardiac patient, the enlarged ECF does not help to compensate for the decreased EfCV. The cardiac output is still poor and activation of renin secretion will proceed, since renal baroreceptors still sense the reduced EfCV.

5. How does a Valsalva maneuver check for normal function of baroreceptors?

Answer The increased intrathoracic pressure during a Valsalva leads to decreased blood pressure which will trigger the expected baroreflex tachycardic compensatory response.

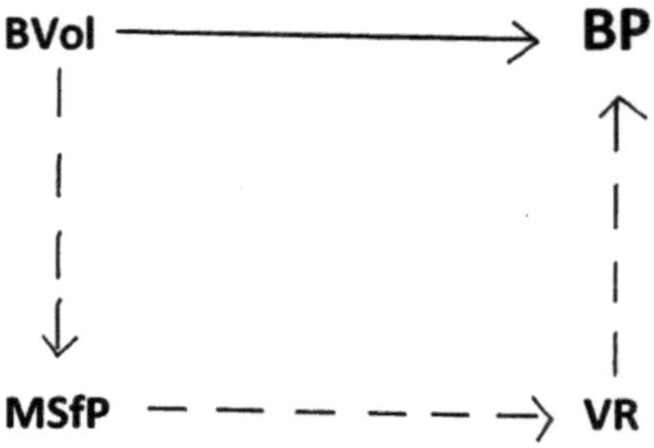

Blood volume determines the blood pressure. Increased blood volume in normal adults will give a more effective circulatory volume to the peripheral tissues. Increased blood volume will increase the venous return by raising the mean systemic (circulatory) filling pressure

Concept The forced expiration against a closed glottis during a Valsalva effort will increase the intrathoracic pressure. This has the effect in elevating the central venous pressure and will reduce the venous return. An acute drop in blood pressure due to a decreased cardiac output will activate the baroreflex. The baroreflex will increase the sympathetic discharge.

The increased sympathetic activity serves to restore the blood pressure by attempting to increase the cardiac output as well as the TPR. A tachycardia is thus observed when the Valsalva maneuver is sustained, demonstrating a normal baroreceptor response.

When the person lets go and abandons the Valvalsa effort, a sudden return of the central venous pressure to normal causes a rebound increase in the venous return. Arterial blood pressure rapidly rises at that point. This time, the baroreceptor will detect the increased vascular pressure stretch and a feedback bradycardia is produced as a result of concurrent reflex increased parasympathetic/decreased sympathetic inputs to the sinoatrial pacemaker node.

6. How does the renal sympathetic nerve (RSN) help to restore blood volume after blood donation?

Answer The action of the RSN conserves total body sodium by decreasing filtered sodium load and increasing renal reabsorption of sodium.

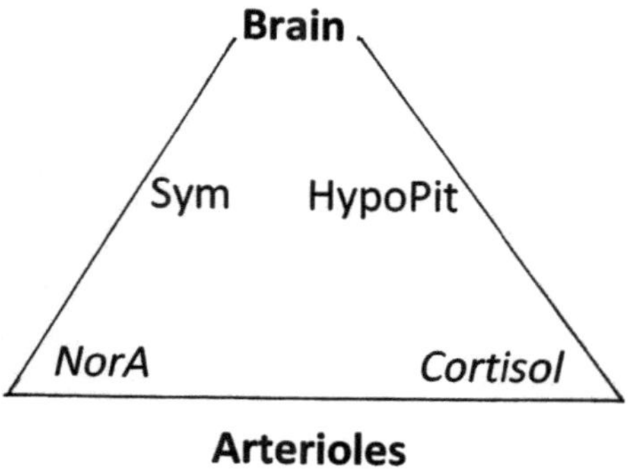

The vascular arterioles function in modulating total peripheral resistance in blood pressure control as well in as local regional flow regulation. Noradrenergic vasoconstrictor sympathetic fibers innervate the arterioles and determine the degree of vascular resistance. The anterior pituitary, under hypothalamic control, regulates adrenal cortisol secretion and this steroid hormone is needed for normal vascular smooth muscle responsiveness to vasoactive agents

Concept The action of the RSN always decreases urinary sodium excretion. Sympathetic is "sympathetic" to sodium balance.

Hypovolemia triggers a baroreflex increase in sympathetic nerve effector actions. This includes the renal sympathetic arm of the autonomic neural activity. Physiologically, the increased RSN action must restore blood volume.

The RSN effects the volume control by its effect in conserving sodium or reducing urinary excretion of sodium. This is accomplished, because the RSN reduces the filtered sodium load. At the same time, the RSN's action in the kidneys results in more sodium reabsorption.

The filtered load effect is through decreasing glomerular filtration rate (GFR) due to sympathetic vasoconstriction of the renal arterioles.

The heightened sodium reabsorption is due to a direct action of sympathetic fibers that innervate the proximal tubular epithelial cells. This segment of the nephron normally reabsorbs ~70 % of the filtered sodium load.

The granular JG cells that secrete renin are innervated by RSN. Thus, the sympathetic increase in plasma renin will result in active pathways that reabsorb sodium. Sodium is recovered more from the tubular fluid by actions of both angiotensin II and aldosterone.

7. How does angiotensin II affect the two determinants of blood pressure equation?

Answer Angiotensin II (AII) stimulates aldosterone secretion from the adrenal glands. AII is also a strong vasoconstrictor that increases the TPR.

Concept Angiotensin II is a multitasker in the physiologic control of blood pressure. As a potent vasoconstrictor, angiotensin II (AII) raises the TPR in the blood pressure equation (BP = cardiac output (CO) × TPR).

Like the sympathetic activity, this vasoconstricting action of AII would be selective to make physiologic sense. In essential organs like the brain and the heart, we can expect AII not to be active in increasing the vascular resistance.

In the kidneys, AII acts to complement renal sympathetic action on the afferent/efferent arterioles. This action temporally reduces the renal blood flow (RBF)/GFR, and this decreases the filtered sodium load. In hypovolemic/hypotensive situations, AII has the effect of reducing sodium excretion in the overall scheme of volume control.

Excreted sodium = filtered sodium minus reabsorbed sodium.

AII increases the renal sodium reabsorption directly by stimulating the function of the proximal tubular cell. Of course, AII is one of the primary stimuli (the other is hyperkalemia) for the release of aldosterone from the adrenal cortex.

Thus, the action of AII not only increases the TPR, but AII is also a key steroid hormone that helps to maintain ECF/blood volume. This latter parameter is a positive factor of cardiac output.

A few other actions of AII also relates to restoring volume during fluid loss. AII stimulates vasopressin secretion from the posterior pituitary and AII is also a dipsogenic.

8. Why does the diastolic blood pressure (DBP) change less than the systolic during exercise?

Answer The decrease in the TPR during exercise consequent from vasodilation in the muscles and the skin tend to minimize the diastolic pressure.

Concept The systolic pressure is determined by two main factors, the stroke volume and the arterial compliance. The diastolic pressure is due to the elastic recoil of the aorta and large arteries, and this provides the driving pressure for continuous blood flow during ventricular relaxation. A major factor that affects the DBP is the TPR (imagine DBP as the elastic recoil of an inflated balloon and the outlet the TPR; if the outlet is more restricted the balloon DBP is higher and vice versa).

During physical activity, an extensive vasodilation in the skeletal muscles has the effect in lowering the TPR. In addition, the need to maintain body temperature by losing heat during exercise is effected by cutaneous vasodilation. The overall TPR is thus reduced. Selective sympathetic constriction in the splanchnic and the renal vasculature helps to maintain an adequate blood pressure to power the greater tissue perfusion to the muscles.

The cardiac sympathetic activity to the heart increases the cardiac output, which also sustains a higher blood pressure in spite of the decreased TPR during exercise.

For a given increased stroke volume, the systolic pressure would be higher. So should the elastic recoil pressure (DBP) from the greater stretch of the aorta by a bigger ejected blood volume. However, the decreased TPR lessens the effect on the DBP by the increased stroke volume.

9. How does the change in capillary blood pressure during blood loss help to compensate for the vascular volume contraction?

Answer Transcapillary shift of fluid from the interstitium into the capillary compensates for the vascular volume contraction as the decreased capillary hydrostatic pressure reduces filtration and tends towards capillary reabsorption.

Concept In the classical capillary described in physiology texts, the net filtration of fluid occurs at the arteriolar end and net reabsorption of fluid at the venular end (the student should note that in some organs, reabsorption takes place along the capillary, e.g., intestines). About 90 % of the filtrate at the microcirculation is reabsorbed, the remaining 10 % recycled by lymphatic drainage in to the systemic circulation.

The capillary dynamics above is dependent on the balance of Starling forces along the capillary and in the interstitium.

During hypovolemia, the baroreflex increase in sympathetic activity increases the arteriolar resistance in many organs including in the skeletal muscles. "Downstream" from the arterioles, the capillary hydrostatic pressure is reduced. This effect on capillary hydrostatic pressure produced by sympathetic arteriolar constriction reduces even more, the drop in capillary pressure from the hypovolemia itself.

The balance of Starling's capillary oncotic and decreased capillary hydrostatic pressure will shift the balance to a net reabsorption that might occur along the length of the capillary from arterioles to venules. (in splanchnic capillary, the usual reabsorption is enhanced by the fall in hydrostatic pressure).

In the kidneys, the glomerular capillary filters less fluid due to the decreased RBF/GFR when the RSN constricts the arterioles.

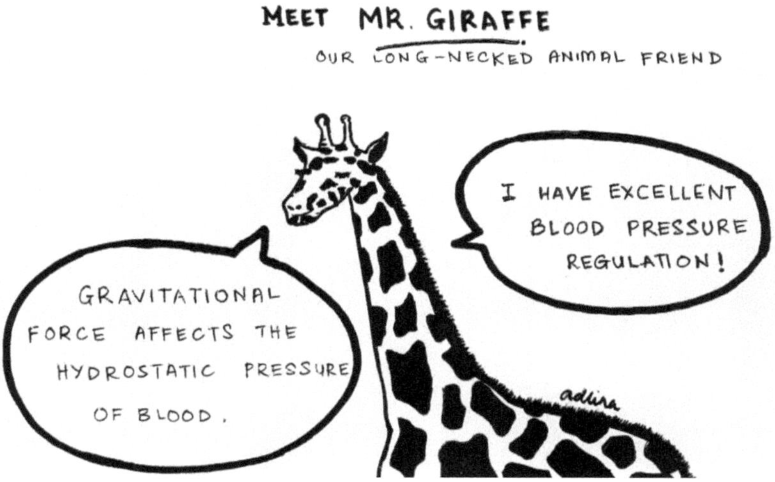

Our heads are above our hearts for most of our human days. One key reasons for maintaining an adequate arterial blood pressure is to ensure sufficient cerebral blood perfusion to our brains. The location of the baro sensors at and above the heart's level is also a creatively appropriate design, to allow the baroreceptors to monitor a decrease in blood pressure. This blood pressure regulation will be appreciated when we next look at a Giraffe!

Thus, the transcapillary shift of fluid into the vascular compartment is one of the diverse compensatory mechanisms for hypovolemia. This is sometimes described as "autotransfusion" of fluid from the interstitium to the blood plasma.

10. How is the perfusion pressure for coronary blood flow affected in aortic stenosis?

Answer The aortic pressure that determines the coronary perfusion pressure is decreased since the aortic pressure is "downstream" from the narrowed aortic valve.

Concept The coronary blood supply to the ventricles is affected by the cardiac muscle contraction. The mechanical contraction compresses the blood vessels during systole. Thus, especially at the left ventricle where a higher muscle tension is generated to produce a higher systolic intraventricular pressure of 120 mmHg, considerably more blood flow occurs during diastole when the ventricle muscles relax.

During systole, due to the aortic stenosis, the aortic blood pressure does not rise as high during ventricular ejection. Therefore, during diastole, the "head" pressure for perfusing the coronary vasculature is also reduced.

When the aortic valve is narrowed, the ventricular muscle tension is heightened during pumping contraction. This also lessens the coronary blood flow due to the greater compression on the coronary vessels.

If the aortic valve fails to close normally, there will be a back-flux of blood into the ventricles. This will also lower the diastolic pressure in the aorta. Again, the diastolic coronary perfusion to the ventricles is reduced.

Chapter 4
Systemic Circulation and Microcirculation

The microcirculation refers to the end-user point of the cardiovascular system. This is the capillary network that supplies the cells with oxygen/nutrients and drains away the metabolite CO_2 and other byproducts of cellular activity. There is a continuous flux and exchange of fluid and solute at the capillary. The function of the lymphatic drainage system is also associated with the capillary dynamics.

1. In general, what proportion of capillary filtrate is reabsorbed back into the capillary at the venule end?

Answer About 90 % of capillary filtrate is generally reabsorbed back at the venular end of the capillary.

Concept In the classical "textbook capillary" capillary filtration is shown to occur at the arteriolar end of the capillary. Towards the exit of the capillary into the venules, reabsorption of fluid takes place. This net filtration and net reabsorption at different ends of the capillary is explained by the progressively decrease in hydrostatic pressure along the capillary (from 30 mmHg to 15 mmhg). The capillary oncotic pressure however is relatively constant along the capillary length (~25 mmHg).

The other two Starling's extracapillary interstitial forces do not change. The interstitial oncotic pressure about 5 mmHg and the interstitial hydrostatic at zero or even negative mmHg. Computing the four Starling's forces indicate a positive net filtration pressure at the arteriolar end and a negative net filtration (positive reabsorption) near the venule.

About 90 % of the filtrate is recycled back into the capillary. The balance 10 % is returned to the systemic blood via the lymphatic drainage. There is thus a major plasma fluid circulation in the blood vessels, powered by the contracting heart and a minor fluid recirculation through the lymphatic system. The lymphatic also drains some plasma proteins that leak out, so the interstitial oncotic pressure does not build up to interfere with capillary reabsorption.

The student should note that in certain organs, the capillary dynamics are quite different as they are adapted to the function they serve. For example, in the intestines, reabsorption of fluid occurs along the capillary length, which is appropriate for the high absorptive activity of the gut.

H. M. Cheng, *Physiology Question-Based Learning*, DOI 10.1007/978-3-319-12790-3_4

Conversely, the glomerular capillary only filters with a significant fraction of 20 % of renal plasma flow entering the Bowman's capsule. The glomerular oncotic pressure thus rises among the glomerulus.

In contrast, for the renal peritubular capillary which is downstream in series from the glomerulus, reabsorption is perhaps the sole capillary event as in the intestines.

2. What is the main determinant of blood viscosity and effective osmotic pressure in the microcirculation? (Red blood cell (Rbc) plasma proteins)

Answer Blood viscosity is determined predominantly by the hematocrit. The osmotic pressure at the capillary is due to the plasma protein which is nonpenetrating at the endothelial cells.

Concept The blood volume is made up of the hematocrit and the plasma volume. Hematorcrit is about 45 % in males and slightly lower in females. Blood viscosity is a factor that contributes to the vascular resistance. Unlike the vascular radius that can be regulated by sympathetic nerve and vasoactive agents (and resistance is inversely related to radius[4]), the hematocrit is homeostatically maintained by controlled erythropoiesis.

In polycythemia, the viscosity and thus the resistance to blood flow is increased. Thus, the compensatory secondary polycythemia from exposure to high altitude hypoxia has a self-limiting benefit. If the increases in blood viscosity begin to make the flow sluggish, then the delivery of oxygenated blood to the cells is still compromised.

The plasma protein concentration provides the effective osmotic pressure at the microcirculation. The vascular space is importantly maintained by this plasma oncotic or colloid osmotic pressure. The value is 25 mmHg, a small value (less than 2 mOsm/L) compared to the total plasma osmotic pressure equivalent to ~300 mOsm/L.

Sodium and its associated anions, however, are freely penetrating at the capillary and do not exert an osmotic pressure to determine the fluid movement at the microcirculation.

The plasma oncotic pressure is the essential osmoactive force that preserves the volume of the vascular compartment.

Sodium and its anionic partners are the predominant osmo-active solutes at all cell membranes and is the deciding factor for fluid shift between the intracellular and extracellular spaces.

For fluid movement between the interstitium and the vasculature, the governing "prince" is protein.

3. How are the hemodynamic determinants of flow in a single capillary applied to the whole systemic circulation from left to right side of the heart?

Answer The flow is equal to the perfusion pressure over the vascular resistance in a single capillary, but in the capillary network, the overall resistance is reduced by the extensive parallel vascular conduits.

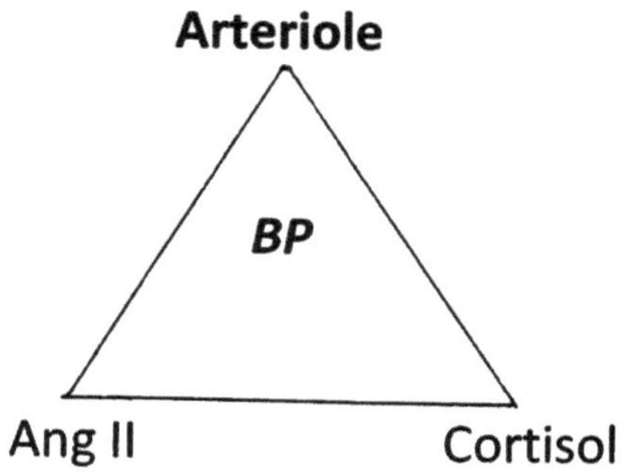

Angiotensin II is a strong vasoconstrictor and elevates blood pressure (*BP*) by increasing the systemic total peripheral resistance. The glucocorticoid cortisol is needed for vascular responsiveness, a permissive action for normal vascular sensitivity to vasoactive agents

Concept The hemodynamics of a blood flow in the vessel is described by the same factors that govern fluid flow in a rigid tube. Flow = pressure difference/resistance to fluid flow.

The resistance is inversely related to the fourth power of the radius in the single blood vessel.

If we, then, compare one capillary with one artery, the arterial resistance would be less if the same pressure gradient and flow rate is present at both blood vessels.

In the body, however, the major resistance to blood flow is not at the capillary section but at the arterioles. This is due to a number of reasons.

Firstly, the capillary network is an extensive branched microcirculation that ensures blood perfusion to all corners of the tissues. This microcirculatory tree lowers the overall capillary resistance to blood flow.

The arterioles are the resistance vessels in the systemic circulation. The arteriolar smooth muscles are innervated by sympathetic vasoconstrictor nerves. There is a basal vasoconstriction vasomotor tone that contributes to the physiologic function of the arterioles as vascular resistance "gate control" for blood perfusion "downstream" into the capillary microcirculation.

When we apply the flow = pressure/resistance to the entire systemic circulation from left to right sides of the heart, the equation converts to

Cardiac output Flow = Arterial Blood Pressure / total Peripheral resistance

For the pulmonary circulation, which is in series with the systemic circuit, the hemodynamics will be

Pulmonary Blood Flow = Pulmonary arterial pressure / pulmonary vascular resistance

4. How does the capillary blood-flow rate (ml/min) compare to the arterial blood flow?

Answer The normal capillary flow rate is the same as the arterial flow rate in the closed circulatory system.

Concept It is quite essential to use cardiovascular language accurately. When we ask "how fast does the blood flow?," we are referring to the velocity of blood flow (unit will be distance/time). However, when we say "blood flow rate," we should be meaning the volume of blood per time.

In a close circulatory loop, the blood flow rate is the same at any segment of the circulatory system of vascular traffic. The two circulations, the systemic and the pulmonary have the same blood flow rate, i.e., the cardiac output and the pulmonary blood flow has the same value; 5 l/min in a 70 kg, male adult.

The velocity of blood flow is inversely associated with the total cross sectional area (CSA) of the vascular segment. The capillary network has the highest CSA and the corresponding lowest velocity of blood flow. The high CSA is obviously for the diffusion of solutes, respiratory gases and filtration of fluid between the microcirculation and the cells.

The slow velocity of capillary blood flow complements the high CSA is allowing optimal time for capillary exchange, the key function of this vascular life-giving network.

If we examine the hemodynamics, the flow in the blood vessel as volume/time can be expanded to CSA × length of vessel/time. This converts to flow = CSA × Velocity. Since the flow (ml/min) is the same in every sector of the circulation, the velocity of blood flow at each stage of the "bloody" journey will be inversely dependent on the total CSA.

5. How does the hemodynamic principle explain the myogenic response at the precapillary arteriole during hypotension?

Answer The myogenic feedback responds by relaxing the arteriolar smooth muscle to produce vasodilation during decreased perfusion pressure.

Concept When blood pressure decreases, the feedback mechanisms from the baroreflex will produce an increased in centrally mediated (brain stem cardiovascular control neurons) sympathetic vasoconstriction.

It should be noted that this brainstem reflex will help to restore the arterial blood pressure centrally via the baroreflex. The additional point to note in this increase in "total" (better word is overall) peripheral resistance does not include and affect the blood flow to essential tissues of the heart and the brain. Sympathetic nerve

Control of sodium balance is part of the extra cellular fluid (ECF)/blood volume control and blood volume is a major determinant of blood pressure

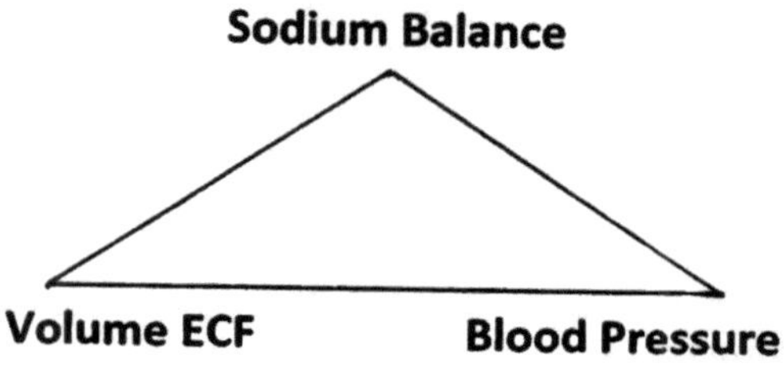

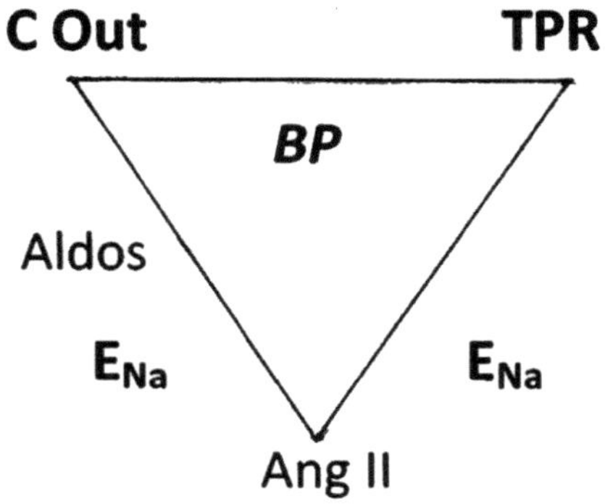

Angiotensin II (*Ang II*)is a multitasker in blood pressure (*BP*) control. The mineralocorticoid Aldosterone is stimulated to increase sodium conservation, and this decreases the excreted sodium. Increased sodium reabsorption helps raise the blood volume, a determinant of cardiac output. The excreted sodium is also reduced by *Ang II* action on renal arterioles to lower the glomerular filtration rate (GFR). *Ang II* also increases the total peripheral resistance (TPR) component that factors into BP

regulation of arteriolar radius is not a key feature in both cerebral and coronary circulation.

In the heart and the brain, the intrinsic autoregulation of blood flow protects the organs from fluctuations in blood perfusion pressure. The baroreflex maintenance of an adequate blood pressure during hypotension is supplemented by local vasodilation of the coronary/cerebral arterioles.

This local, flow autoregulatory response is an inherent property of the arteriolar smooth muscle, involving either less or more calcium influx to produce vasodilation or vasoconstriction, respectively. This is known as the "myogenic response".

Besides the "myogenic" mechanism, a local metabolite vasodilator feedback is also operational during hypotension to compensate for the reduced blood flow in the cerebral/coronary vasculatures.

6. How is venous pooling linked to changes in capillary dynamics? (edema) (no arterial pooling, arteriolar resistance)

Answer Venous pooling increases the venous pressure and the capillary hydrostatic pressure as well.

Concept When a person stands stationary without voluntarily moving his calf muscles, the gravitational force will exert an increase in the hydrostatic pressure in his blood vessels below the level of his heart.

There will be a progressive increase in both arterial and venous pressure, at 0.77 mmHg higher for every cm below heart's level. The arterial pressure in a large artery at the foot (105 cm below the heart) is 180 mmHg [$100+(0.77\times 105)$].

However, the gravitational effect of increased venous pressure affects the capillary pressure more than the rise in the arterial pressure. This is because the resistance of the precapillary arterioles is much greater than the low venular postcapillary resistance.

The higher compliance of the veins produces the increased venous volume during venous pooling. There is no equivalent of arterial pooling due to arterial elasticity that opposes the stretch. The same increase in vascular transmural pressure distends to a greater degree the more compliant veins.

This increased venous/capillary hydrostatic pressure will tend to produce a peripheral edema of the dependent areas of the lower limbs. We may know from experience that our ankles feel more tight and heavy when we have been sitting long with little mobility, e.g., on a long air flight.

7. What are the two major drainage functions of lymphatics that maintain the optimal interstitial space?

Answer The lymphatic drainage returns both unreabsorbed capillary filtrate and plasma proteins that leaked out of the capillaries.

Concept The total interstitial fluid volume is three times the plasma volume. The capillaries constantly filter plasma fluid minus the plasma proteins. There is a local fluid recycling as most of the filtered fluid is reabsorbed downstream along the capillary.

The ~10% of fluid is recycled back into the systemic circulation via the lymphatic vessels. The net capillary filtration rate is equal to the rate of lymphatic fluid flow. There is no accumulation of excess fluid in the interstitial space.

If fluid builds up in the interstitium due to a disturbance of capillary/lymphatic dynamics, edema is the resulting named condition. Edema increases the diffusion distance and interferes with capillary exchange. Peripheral edema is relatively noncritical compared with the case that the edema occurs in the lungs (pulmonary edema), when lung oxygenation will be reduced.

The lymphatics also maintain the interstitial oncotic pressure at a minimum mmHg. Some plasma proteins do leak and trickle out of the microcirculation. These proteins are also drained through the lymphatics and added back into the systemic blood circulation. Should the oncotic pressure in the interstitium increase, this can also potentiate the development of edema.

Impaired lymphatic drainage occurs in different situations, e.g., surgical removal for malignancy, in filariasis when the parasites infect the lymphatics. Lack of muscle activity also reduces the compression of lymph vessels and lymphatic circulation, as it occurs in an obedient soldier standing at attention.

The lymphatic vessels have one-way valves for the unidirectional flow of fluid and proteins. The smaller lymphatic vessels merge into larger lymphatics, and finally into the largest lymph conduit, the thoracic duct, that empties into the large veins. Lymphatics also have contractile smooth muscle wall that helps to propel lymphatic flow. This "myogenic" lymphatic "pump" is further enhanced by innervation from the excitatory sympathetic nerve.

8. What is the value of the lymphatic flow per day?

Answer About 3 l/day.

Concept Knowing some quantitative aspects of physiology is very helpful in appreciating the homeostatic dynamics in the human body. Functionally, there are four physiologic circulations namely the systemic, the pulmonary, the capillary microcirculation and the lymphatic flow.

Systemic circulation has a value per day of (5 l $\times$ 60 $\times$ 24) 7200 l of blood flow daily. This would be the same for the pulmonary circulation since the left and the right ventricular pumps are arranged in series in the closed circulatory loop of the entire cardiovascular system.

At the capillaries, the microcirculation refers to the general capillary filtration and capillary reabsorption circuit. This capillary dynamics serves as the essential function of capillary exchange at the tissues. On average, 20 l of plasma fluid per day are filtered. As a percentage of total plasma flow per day, this works out to be just a small percentage of less than 1 %. Total plasma flow is ~60 % of 7200 l/day which gives 4320 l/day. The capillary filtration is 20/4320, which is 0.46 %.

Normally, about 90 % of capillary filtrate is reabsorbed into the venular end of the capillaries. The balance 10 % of excess fluid is flushed out by the lymphatic drainage and returned to the systemic circulation. An edema from an accumulation of fluid in the interstitial space is prevented. Of the 20 l of plasma fluid filtered/day, about 3 l of fluid flow back via the lymphatic vessels.

Three liters might not seem like a lot of fluid volume. In a 70 kg male adult with a total body water of 60 % of body weight at 42 l, around 15 % of the total body water is in the interstitium, i.e., 0.15 $\times$ 42 l giving 6.3 l.

When we work out this value for interstitial fluid (ISF), it is clear that 3 l of daily lymph fluid represent almost half of the total ISF volume. Obviously, an edema easily develops if the lymphatic circulation is reduced.

The student should note that the normal blood volume in the 70 kg physiological male standard is 5 l. The 3 l of lymphatic flow/day is the entire plasma volume!

Four cheers to the four integrated physiologic circulations!

9. How is capillary dynamics different in the GI tract compared to skeletal muscle

Answer The GI tract has both a variety of digestive secretory and absorptive functions. Both these events will involve capillaries and a different balance of the capillary exchange Starling's forces.

Concept The secretion of digestive juice by the epithelial cells of glands along the GI tract involves fluid movement from the capillarized gland into the lumen and eventual empyting into the GI tract. The aqueous medium of all secretions is derived from the blood capillaries that supply the glands. Osmotic gradient generated by prior solute secretion drives the water movement during the secretion. The net Starling's forces along these capillaries must be suited to favour filtration.

Conversely, intestinal absorption of water also follows solute absorption that creates the local osmotic difference between the lumen and the interstitium. Water is then absorbed both trans- and paracellularly into the capillaries.

Thus, for this absorptive function to be effective, the balance of the Starling's forces along the length of the capillary should favour fluid reabsorption.

In the pathophysiologic case of "dumping syndrome" due to excess hyperosmotic food in the intestine, an abnormal secretion of fluid occurs into the lumen (hypovolemia results)

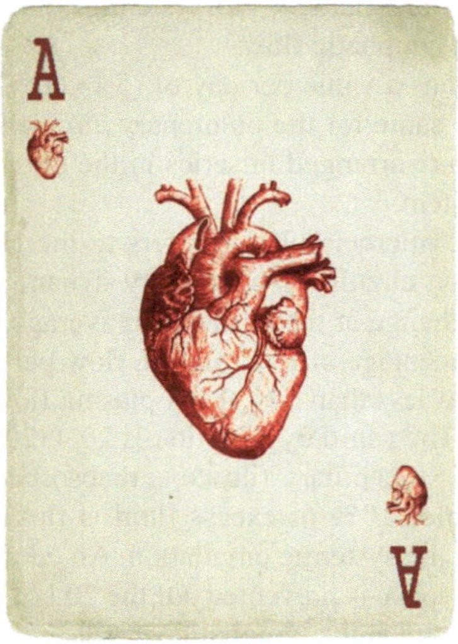

CARDology! We can think of several As essential in cardiovascular functions. Autonomic nervous system, adrenaline, angiotensin II, atrial natriuretic peptide, arteriolar radius regulation, adrenal glucocorticoids, antidiuretic hormone

In skeletal muscles, the normal profile of capillary filtration at one end and reabsorption at the other end downstream is altered, when the blood volume and blood pressure falls. The capillary hydrostatic pressure is decreased by the systemic hypotension and also by the reflex sympathetic vasoconstriciton of the precapillary arterioles.

This reduction in capillary blood pressure changes the profile of the Starling's forces, and absorption of fluid along the length of the capillary is observed. This inwards transcapillary shift of fluid is part of the compensatory events that helps to maintain the vascular volume during hypovolemia.

10. Why are the delicate thin walled capillaries not easily prone to rupture?

Answer The relative insusceptibility of the capillaries to rupture is due the operation of Laplace's Law that explains the protective effect of the small radius of the capillaries.

Concept The physical principle explained by the law of Laplace states that the tension in the wall of a cylinder (T) is equal to the product of the transmural or distending pressure (P) and the radius divided by the wall thickness (w).

$$\text{Tension} = \text{P} \times \text{radius}/\text{thickness}$$

Since the interstitial pressure is low, the trasnmural pressure (pressure difference across the wall) is basically the intraviscus pressure. If the viscus is thin walled, the w factor is also insignificant and can be ignored.

In a cylinder such as a blood vessel, Laplace's equation becomes $P = T / \text{radius}$.

As such, the smaller the radius of the blood vessel, the less is the tension in the wall that is required to balance the distending pressure. For example, in the human aorta, the tension at normal pressures is about 170 kdynes/cm and in the vena cava, it is about 21 kdynes/cm.

In contrast in the capillaries, it is only 16 dynes/cm.

This law of Laplace also explains the problem associated with a dilated heart. When the radius of the ventricular chamber is increased, a greater tension must be developed in the myocardium in order to generate the same intraventricular pressure. The failing heart is weakened further by the increased cardiac work needed.

In the lungs, the Laplace's law also accounts for the observation that in the absence of surfactant, the smaller alveoli tends to collapse under the effect of the alveolar surface tension.

Pee wee! Laplace also has something to say about our urinary bladder. As the bladder is filled, the tension increases. The first urge to void is felt when the bladder volume is about 150 ml. The next marked sense to void a full bladder happens at about 400 ml. The small change in intrabladder pressure between the two volumes is due to the increasing radius that balances the increasing wall tension as the bladder fills.

Chapter 5
Regional Local Flow Regulation

"*Ask it shall be given.*" The heart and the peripheral organs it supplies is like a mother and her children. The mother heart is the source of all sustenance, its oxygenated and energy substrate-richblood (cardiac output). Some of the organs are more demanding;at rest, more blood flow/weight of the organ. There is anunequal distribution of cardiac outputto the different organs. In specific situations,there is a redistribution of cardiac output based on the differential functional needs of the organs. In general, a greater demand for oxygen/energy by a particular organ will lead to an increased blood flow to that tissue. The mother heart pampers all her children and gives more of its generous life-giving output to whichever organ that ask for more! Let us look at some scenarios in the body that illustrates this cardiovascular "ask and it shall be given" principle.

1. How does the heart muscle ensure that it receives more blood when it is more active?

Answer Increased cardiac metabolism releases vasodilator metabolites that provide the greater coronary perfusion commensurate with the needs of more active heart pump.

Concept The cardiac muscle has its own self-regulating intrinsic mechanisms that ensure that the myocardium has adequate blood perfusion for its essential function. At rest, when the blood pressure fluctuates over a certain physiological range, autoregulatory mechanisms maintain a relatively constant coronary flow.

When the heart becomes active during physical activity (or by emotional stimuli), the heart pumps faster and more strongly under the action of both cardiac sympathetic nerve and circulating adrenaline.

Coronary blood flow increases in tandem with the higher cardiac metabolic needs when a bigger cardiac output is ejected. The increased coronary flow, however, is not due to any direct sympathetic vasodilator nerve. Adrenaline might dilate some vessels by its action on any vascular beta adrenergic receptors.

H. M. Cheng, *Physiology Question-Based Learning*, DOI 10.1007/978-3-319-12790-3_5

The coronary vasodilation is primarily due to increased local metabolite vasodilators in the myocardium. The cardiac sympathetic action by its tachycardic and inotropic actions increases the cardiac metabolism. Thus, the effect of sympathetic action in increasing the coronary perfusion is indirect via the metabolite vasodilators. One such established vasodilator is the adenosine tri-phosphate (ATP) metabolite, adenosine. This phenomenon of metabolic-driven perfusion of the myocardium is also termed "active hyperemia." There is also "reactive hyperemia" resulting from the compression of coronary vasculature especially by the strong left ventricular muscle contraction.

2. At rest, which organ is the most "demanding" and receives the highest rate of blood flow?

Answer The pulmonary blood flow is the cardiac output from the right "heart" (ventricle).

Concept The pulmonary blood circulationis in series with the systemic circulation. Therefore, the cardiac output, defined as from each ventricle (both ventricular pumps in series) is the same from each side of the twin-engine heart.

Changes in pulmonary blood flow during physical activity are obviously not for the purpose of supplying more oxygen to the lungs! (Bringing coal to Newcastle, or in Malaysia we might say,"Bringing palm oil to Kuala Lumpur!")

The circulatory purpose of increased pulmonary blood flow during exercise is to deliver a higher rate of oxygen to the cells (ml oxygen per time). This is partly due to more lung oxygenation. However, since maximum oxygenations is "perfusion-limited" (see Part 2 "Respiratory Physiology"), the greater pulmonary blood flow is critical for a higher rate of O_2 supply to the tissues.

Vasodilation of the pulmonary vessels during increased pulmonary flow is not due to neuralor metabolite actions. (Hypoxia actually produces a unique vasoconstriction in the lungs; see Part 2 "Respiratory Physiology"). Pulmonary vascular resistance is decreased duringexerciseassociated with increased pulmonary blood flow. The greater right ventricular cardiac output directly distends the pulmonary vasculature in what could be termed "mechanical vasodilation."This is a major vascular event when the pulmonary arterial pressure is higher as the pulmonary vessels are relatively compliant.

3. How does the brain maintain its blood supply when there is a drop of 30 mmHg in arterial blood pressure?

Answer Cerebral autoregulation of flow in conjunction with baroreflex support of arterial blood pressure maintains the blood supply to the brain.

Concept The intrinsic autoregulation of blood flow is operative over the range between 60 and160 mmHg. This indicates that if the blood pressure rises above or drops below 100 mmHg,the cerebral perfusion is autoregulated to sustain a relatively unchanged blood flow.

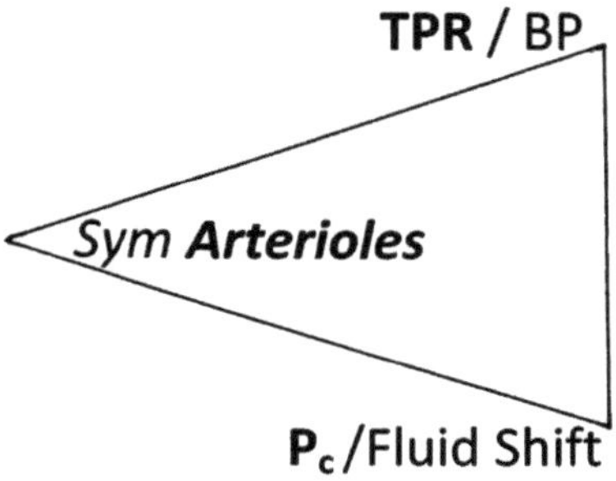

Regulatory sympathetic neural actions on vascular arterioles have a dual compensatory effect. The arteriolar vasoconstriction increases the total peripheral resistance, a determinant of blood pressure. The "downstream" effect of the vasoconstriction on decreasing capillary hydrostatic pressure favors a reverse transcapillary shift of fluid to restore blood volume

In reality, two reflex mechanisms are activated in concert to preserve the cerebral perfusion. Centrally, the baroreceptors will detect the hypotension and trigger cardiac and vasculararteriolar constriction to raise the decreased blood pressure.

At the same time, at the local brain tissues, the cerebral arterioles will vasodilate. This cerebral vasodilation cooperates with the baroreflex action on the blood pressure, and the brain continues to receive relatively normal blood flow.

The cerebral autoregulatory response is mediated by similar mechanisms to the autoregulation at the coronary circulation. One is the myogenic arteriolar response. The smooth muscle of the cerebral arterioles vasodilates when it senses less stretch by the hypotension.Secondly, a metabolic mechanism comes into play.Initial decrease in cerebral flow, before autoregulation kicks in, causes a build-up of vasodilator metabolites. Local increase in PCO_2, localhypoxia and lower pH in the brain tissue also contribute.

There is a marginal ~20% decrease in cerebral blood flow in the presence of autoregulation. A third physical mechanism maintains the cerebral perfusion pressure. The arterial level at the head level drops about 30 mmHg, but the jugular venous pressure decreases 5–8 mmHg. The drop in perfusion pressure is reduced.

Because the cranial volume is fixed, a decrease in the venous pressure lowers the intracranial pressure. The latter decreases the external pressure on the cerebral vessels, and the cerebral vascular resistance is lessened.

4. What is the main determinant of blood flow to the skeletal muscle at rest and during exercise?

Answer At rest, the sympathetic vasoconstrictor innervations set the muscular blood flow. During exercise, the predominant regulator of increased muscle flow is local vasodilating conditions in the muscles.

Concept As in almost all tissues, the arterioles that control inflow of blood in to the capillaries of the organs are innervated by the sole sympathetic vasoconstrictor adrenergic nerves.

At rest, this neural basal input sets the baseline for skeletal muscle flow.

The blood flow increases markedly during physical activity. Is this due to a withdrawal of sympathetic vasomotortone as in the baroreflex response tohypertension?

Although the blood pressure is higher during exercise, there is no reflex bradyardia. Obviously, the exercise tachycardia is the natural expected event from stimulation by cardiac sympathetic nerve. We can expect the same for the sympathetic nerve activity to the arterioles of the muscles. There appears to be a resetting of the baroreflexso that in the exercise hypertension and tachycardia can sustain the increased cardiac output and blood flow to the muscles.

Are there then vasodilator nerves to the muscles? In some mammals, the cholinergic sympathetic nerves have been described that seem to be involved in the initial increase in exercise (or even pre-exercise expectation) muscle blood flow.

When muscle contractions are in progress, local changes in the tissue environment feedback to increase the blood flow. These include local hyperkalemia (from more action potentials), hypoxia, increased PCO_2, and lower pH (lactic acisosis).

The student should not mix up the increase in outflow venous return due to the "musclepump" with the increase in inflow skeletal muscle flow that is effected by the diverse local vasodilatory agents.

There is a decrease in total (overall)peripheral resistance during exercise due to the vasodilation in the muscles (cutaneous vasodilation also to lose heat).

5. How and why does the blood flow increase to the skin during a hot day?

Answer Cutaneous vasodilation is to lose heat in order to maintainnormal body temperature.

Concept Body temperature is controlled by the hypothalamus thermoregulatory neurons. The thermostatic set point of normal body temperature 37 °C is regulated by either heatloss mechanisms of heat consecrating responses. The hypothalamicthermocontrol neurons receive afferent inputs from both central and skin thermoreceptors.

During exposure to hot environment, the heat loss is promoted. One major way is cutaneous vasodilation. This results from a decrease in the activity of sympathetic vasoconstrictor fibers to the skin blood vessels.These autonomic nerves are adrenergic.

At the same time, sweating is triggered. Sweat glands are also innervated bysympathetic nerves that release acetylcholine as neurotransmitter. Thus, a rise in body temperature sets in action a concurrent reduction in vasoconstrictor sympathetic nerve to vasodilate the cutaneous vessels and an increased cholinergic sympathetic stimulation of sweating.

When the sweat is vaporized from the skin surface, the latent heat of vaporization is used, and this heat extracted from the body then cools the body temperature.

A different situation occurswhen a person is frightened and begins to have cold clammy skin ("cold sweat"). The coldness is due to increased sympathetic vasoconstriction of the cutaneous blood vessels. Sweating is stimulated by the separate cholinergic, sympathetic neural pathway.

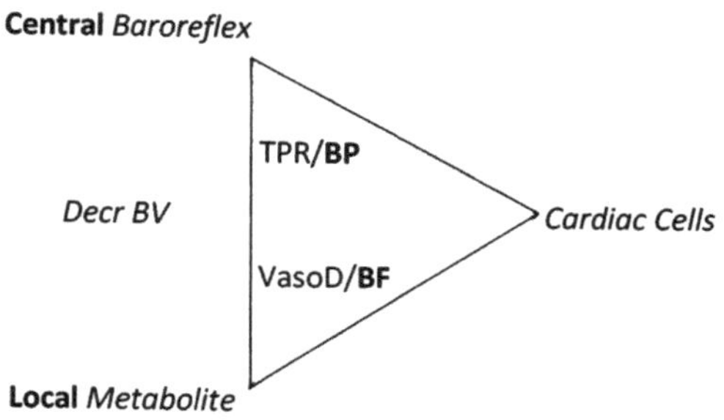

The regulation of normal flow (*BF*) to the heart involves integrated central baroreflex vasoconstriction to maintain adequate systemic perfusion pressure operating in concert with local coronary vasodilation. This occurs, e.g., during hypovolemia. In exercise hyperemia, selective vasoconstriction occurs in splanchnic/renal vasculature with vasodilation in the skeletal muscles and the skin (to lose heat) besides increased coronary perfusion

6. What is the physiologic rationale for a decrease in renal blood flow during hypovolemia?

Answer Reduction in renal blood flow during hypovolemia is part of the compensation to restore blood pressure and blood volume.

Concept At rest, about 20 % of cardiac output is distributed to both kidneys. This is a relatively large regional blood flow with respect to the mass of the renal tissues. Besides supplying oxygenated blood and energy substrates to the cells of the kidneys, the renal blood flow has the primary function in channeling blood to the nephrons for filtration.

Compared to the 1 % of plasma filtered in other capillaries, the glomerular capillary has a filtration fraction (glomerular filtration rate, GFR/renal plasma flow) of 20 %.

During hypovolemia, the blood pressure is also decreased. The kidneys have a key role in blood volume control, and this is linked to its homeostatic function in blood pressure regulation. The students need to review and appreciate that extracellular fluid (ECF)/blood volume control is integrated into the renal handling of sodium and regulation of sodium balance.

Hypovolemia triggers a general increase in sympathetic discharge and this includes increased renal sympathetic nerve (RSN) action. The RSN's overall action in the kidneys is to conserve sodium together with effects to restore volume and blood pressure.

Vasoconstriction of renal arterioles by RSNis part of the compensatory increased total peripheral resistance that is produced by greater sympathetic neural activity.

The temporal decrease in GFR consequent from the reduced renal blood flow is to decrease the excretion of sodium and conserve body sodium to recover the volume. Filtered load of sodium is reduced when the GFR drops.

The kidneys excrete less sodiumalso because the RSN acts to increase the tubular reabsorption ofsodium. This is via stimulating the release of renin from the juxtaglomerular endocrine cells of the afferent arteriole. RSN also has direct action on proximal tubular sodium reabsorption.

7. What is vasovagal shock?

Answer This is a neurogenic shock when a sudden change, burst, or outflow of autonomic neural activity produces vasodilation, pooling of bloodin the extremities, and the person faints.

Concept Vasovagal, neurogenic shock is one of three types of what is categorized as distributive shock (Ganong, 2nd edition, p. 637). The other two are anaphylaxis and sepsis. Distributive shock is also called "warm shock," because the skin is not cold and clammyas seen in hypovolemic shock. In distributive shock, the blood volume is normal, but there is amarked sudden vasodilation.

Fainting is, of course, non-life threatening and only short-lived. The horizontal position postfainting is in a real physiologic sense a "homeostatic mechanism," because the supine person will have improved venous return, cardiac output, and cerebral perfusion.

Other triggers ofsyncope are: postural syncope in patients with orthostatic hypotension. Micturition syncope, fainting during urination that is due to a combination of orthostasis and reflex bradycardia activated by voiding urine in these patients (voiding can belike a mild Valsalva!). Cough syncopetakes place when the forced expiratory coughingincreases the intrathoracic pressure and impede venous return.

The carotid sinus syncope is produced when a tight collar presses on the carotid baroreceptors and produces the reflex bradycardia and vasodilation.

In neurogenic vasovagal shock, there would be a burst of vagal parasympathetic to the cardiac pacemaker sinoatrial (SA) node (bradycardia) and a decreased in sympathetic vasoconstrictor activity to the arteriolar vessels (vasdodilation). There could be also an acute outflow of vasodilator sympathetic pathways. Vasovagal syncope occurs in situations of intense emotional stimuli.

The syncope can be due to serious underlying abnormalities, and all cases of syncope must be investigated. About 25 % of syncopal episodes are due to cardiac causes, e.g., transient intracardiac flow, cardiac arrhythmias that acutelydecrease the cardiac output.

Since here, there is an overlap with cardiogenic shock that is described also as neurocardiogenic shock.

The inquiring students may ask "Why is the vasodilationin distributive shock an adverse event? Would not the blood flow to the tissue be better?" Here the overall hemodynamics of the cardiovascular system needs to be recalled.

The arterial blood pressure is determined by the cardiac output (CO) and the total peripheral resistance (TPR). The blood pressure must be maintained for continuous, adequate effective circulating blood volume to the tissues. A reduction in the CO determinant happens in cardiogenic shock. Asudden drop in the TPR will cause the blood pressure to fall. The brain is deprived of sufficient cerebral perfusionand the person faints.

8. How is anaphylactic shock and septic shock similar in their pathophysiology of circulatory failure?

Answer Both anaphylactic and septic shockare conditions classified under vasogenic circulatory shock.

Concept Anaphylactic shock results from a severe allergic hyperreaction to a culprit antigen that the person has been previously sensitized. The antigen–antibody allergic reaction results in the release of histamine from some white cells. Circulating histamine produces the widespread vasodilation of arterioles and abnormal increase in capillary permeability.

In septic shock, a serious infective condition due usually to gram-negative bacteria leads to features of vasogenic and hypovolemic shock. Endotoxins, the cell wall polysaccharides act to cause systemic vasodilation. The capillary permeability is also disrupted, and the plasma fluid enters the interstitial space with the reduction of plasma protein oncotic pressure.

The vascular volume is contracted, and this hypovolemia worsens the initial vasogenic, septic circulatory shock.

A similar vasogenic/hypovolemic shock is seen in dengue hemorrhagic shock. The pathogenesis of dengue viral infection involves the release of many chemical mediators that are part of the body's immune response. Some of these cytokines disturb the capillary functional integrity. Plasmafluid and proteins leak out easily. The dengue picture at this stage is of a person bleeding into his interstitium. There is hypovolemia as a result.

In irreversible circulatory shock, compensatory mechanismsfail to restore the normal blood pressure. Permanent impairment of the blood pressure maintenance machinery has occurred.

If severe hypovolemia was the precipitating cause, ischemic damage to the intestinal tissues can permitthe enteric microflora to enter the systemic circulation.

Some of these commensal bacteria that are now in the blood circulation can be the cause of a vasogenic shock.

9. How does the concurrent stimulation of sympathetic nerve to the heart and to specific blood vesselshelp to redistribute cardiac output?

Answer Cardiac output is increased by sympathetic stimulation. Selective sympathetic vasoconstriction at the regional organs redistributes the increased cardiac output during exercise.

Concept The regional, local circulation to the individual organs are generally arranged inparallel to each other. This allows specific local control of flow without affecting the perfusion to other organs which will be the case if the organs are linked by the blood circulationin series with each other.

During physical activity, the cardiac function is increased by sympathetic stimulation with a higherheart rate and stroke volume. The skeletal muscles require more of the greater cardiac output. There is a channeling of more blood flow to the muscles with a relative reduction to the splanchnic and the renal circulations. This illustrates the selective vasoconstriction at the renal and the splanchnic by sympathetic neural action.

In the hypovolemic situation, there is also selective sympathetic vascular effects. This again includes a decrease in the renal, splanchnic, muscle, and the cutaneous

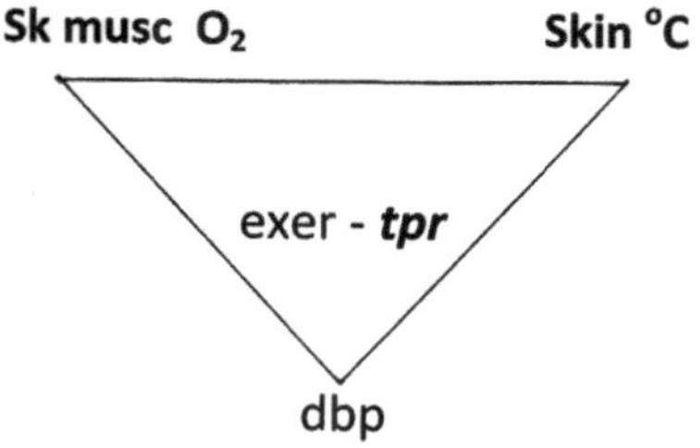

During exercise, heat loss mechanism involves cutaneous vasodilation. Increased muscle activity produces vasodilator metabolites and local hypoxia to increase skeletal muscle blood flow. The overall total peripheral resistance (*tpr*) is decreased. This reduced tpr has the effect of lowering the diastolic blood pressure which is due to the arterial vascular recoil

perfusion. The latter accounts for the "cold clammy skin" during hypovolemia (clammy due to sympathetic nerve to sweat glands).

In exercise, to regulate body temperature, the cutaneous arterioles are vasodilated by a decrease in sympathetic vasoconstrictor activity. Also, during muscle contraction, the actions of local vasodilating conditions/metabolites predominate over any increased sympathetic nerve input to the skeletal muscles.

In hypovolemia, the selective sympathetic action results in an elevated total peripheral resistance in order to restore blood pressure.

In contrast, the selective vasoconstriction at the renal and splanchnic vasculature is exceeded by the vasodilation at the muscles and the skin. Overall, the TPR is decreased. The higher blood pressure in exercise is sustained by the cardiac sympathetic actions.

10. How does severe diarrhea and burn shockaffect local blood flow?

Answer Both severe diarrheaand burn shock produce hypovolemic shock. Diarrhoea results in an isotonic contraction of the ECF and in burns, there is hemoconcentration since the plasma is lost from the burned surfaces.

Concept Hypovolemic shock includes, depending on the cause, hemorrhagic shock, traumatic shock (where blood bleeds into injured muscle/bone tissues), and the loss of fluids due to vomiting or diarrhea.

In hemorrhagic and traumatic shock, whole blood transfusion replaces the vascular volume loss. Saline can also be used. However, saline is distributed into the whole ECF, and thus, only a quarter of the saline volume replenishes the vascular space.

In burn shock where there is hemoconcentration, plasma transfusion is the treatment of choice. There are also "plasma expanders," that are solutions of high-molecular weight sugars that remain in the vascular compartment and actosmoactively.

In hypovolemic, and other kinds of circulatory shock, insufficient perfusion of the tissues leads to increased anaerobic glycolysis. If the resulting lactic acidosis is severe, the myocardial functions can be depressed, and there can be also a reducedvascular responsiveness to vasoactive agents like catecholamines.The acidosis also suppresses the neuronal function, and an acidotic coma is a potential terminal outcome.

Cardiovascular Physiology Ought to be Taught before Respiratory Physiology

This sequential Cardio, then Respi lectures is not merely a matter of departmental preference but is hinged on foundational knowledge building blocks.

A list of rationale for teaching CVS first before Lung Physiology includes

1. General systemic hemodyamics should be taught to students before they are led to consider the peculiar functional characteristics of the pulmonary vasculature.
2. Discussion of Starling's mechanism of the heart to equalize the left ventricular and right ventricular cardiac outputs corresponding to the systemic blood flow and pulmonary blood flow respectively.
3. The Flow, Perfusion pressure and vascular Resistance relationships in systemic flow and pulmonary flow are similar in concept but varies significantly in actual quantitative values.
4. In the systemic circulation, the [Pressure = Flow x Resistance] equation is in most cases interpreted and understood in a right to left direction, Flow and Resistance being Pressure Determinants. However in the pulmonary circulation, the pulmonary Pressure is a determinant of pulmonary vascular resistance (left to right direction) due to the relatively compliant pulmonary blood vessels.
5. All regional arteriolar control of blood flow in systemic circulation respond to hypoxia by vasodilation but in the lungs, there is the special hypoxic pulmonary vaso-constriction.
6. Systemic total peripheral resistance (TPR) does not vary during a respiratory cycle. However, the pulmonary vascular resistance fluctuates during normal inspiration and expiration.
7. Effect of gravity on hydrostatic arterial and venous blood pressure in the standing person. This effect of gravity when understood can then be applied to sorting out the complex ventilation perfusion matching between the apical and basal regions of the upright lungs.
8. The rate of oxygen delivered to the cells is a product of the lung oxygenation (blood oxygen content, ml O_2/ml) and the systemic cardiac output. Control of cardiac output needs to be elaborated to appreciate the cardio-respiratory integrated function.
9. Mechanism of 'respiratory pump' effect during exercise can only be understood if students have prior understanding of central venous/right atrial pressure, venous return.
10. Understanding mechanisms of how a forced expiratory Valsalva manoeuvre affects venous return, stroke volume, blood pressure with compensatory baro-reflex tachycardia requires a foreknowledge o f cardiovascular physiology.
11. Capillary dynamics and balance of Starling 's forces in systemic circulation are disturbed during peripheral edema when the right heart fails. Similar physiologic principles govern the development of pulmonary edema when the left ventricles weaken.
12. Definition of vascular transmural pressure, in particular during blood pooling in the compliant high capacitance systemic veins. Trans-mural pressure is similarly considered in the lungs for airway resistance, trans-pulmonary pressure and pulmonary vascular resistance in the alveolar and extra-alveolar blood vessels.

Do not put the cart before the horse

Hypovolemia is a potent stimuli for adrenal catecholamine secretion. Circulating catecholamines stimulate the reticular activating system in the brain and produce the restlessness observed in some patients with hemorrhagic shock. The restlessness has some homeostatic role since the increased muscle activity and increased respiratory movements provide the muscle and pulmonary "pump" effects to improve the venous return.

Part II
Respiratory Physiology

Introduction: Take a Slow, Deep Breath and Inspire the Concepts

Thinking through physiology to understand well takes time. Work through carefully these questions. Verbalize, (Ventilate!) or better, write down specifically as if you were explaining to another student your answers.

Some dominant concepts embedded in these questions include:

1. How air is inspired or enters into your lungs during unconscious, involuntary breathing.
2. Aerodynamic factors that influence airflow.
3. Lung volumes/capacities, dead spaces, and alveolar ventilation.
4. Parameters for oxygenation at the alveolar–capillary membrane.
5. Unique features of pulmonary vasculature and pulmonary blood flow/right heart cardiac output.
6. The importance of matching or balance between alveolar ventilation and pulmonary blood flow (ventilation/perfusion ratio).
7. Transport of oxygen and carbon dioxide in blood.
8. Neural and chemical regulation of breathing.
9. Cardiorespiratory integrated functions in supplying cellular metabolic needs.
10. Respiratory function and blood pH regulation.
11. Various pathophysiologic conditions of hypoxia and hypercapnia.

In learning respiratory physiology, it is essential to think beyond or outside the "thoracic box." Keep in mind that the alveolar ventilation maintains the partial pressure of O_2 and CO_2 in the alveolar air. The ultimate consumer need is at the cells and cardiovascular, pulmonary blood flow is crucial to meet the end-user tissue requirements. Mixed venous blood in pulmonary artery needs to be reoxygenated and recirculated to the periphery. Think cardiorespiratory physiology!

In pH regulation, the lungs and the kidneys function in concert to handle the carbonic and non-carbonic acid loads added daily to the body. Think Respi–Renal physiology!

AII? Respi–endo

Chapter 6
Airflow

An event taken for granted is the automatic inflow (inspiration) and outflow (expiration) of air from our lungs. Only when the airflow is disturbed, we become conscious of the increased effort of breathing. What drives inspiration and expiration, and what airway parameters affect airflow? How much of the new inspired air reaches the final site of respiratory exchange at the pulmonary alveoli? Do you not think of your breath? Think through these questions.

1. What is the meaning of "negative" in the term "negative-pressure breathing" during normal respiration?

Answer The "negative" refers to the subatmospheric intraalveolar pressure with the atmospheric reference pressure given the value 0 mmHg.

Concept Airflow only takes place when there is a difference in pressure. At the beginning of an inspired breath, the atmospheric and alveolar air pressure is the same, and no air movement occurs yet. Air moves into the lungs when the alveolar air pressure drops to less than the atmospheric pressure (since you cannot change the external pressure!). The respiratory mechanics basically produce this "negative" alveolar pressure to initiate the inflow of air (tidal volume) in to the lungs. Inspiration is an active process and begins with the contraction of inspiratory muscle (diaphragm being the main muscle). This contractile action makes the intrapleural pressure (Pip, between the visceral and pleural membranes) more negative. The change in the Pip then increases the transpulmonary pressure (difference between alveolar air and the Pip). The greater transpulmonary pressure begins to distend the alveoli and the intraalveolar pressure decreases to below the atmospheric pressure (below 0 mmHg of 760 mmHg).

2. What prevents the lungs from collapsing even with the maximal expiratory effort?

Answer The opposing outward recoil of the chest wall to the inward recoil of the lungs.

Concept The lungs are expanded from birth (when we cry out and make our first voice to the world!).The lungs and the chest wall are anatomically apposed

H. M. Cheng, *Physiology Question-Based Learning*, DOI 10.1007/978-3-319-12790-3_6

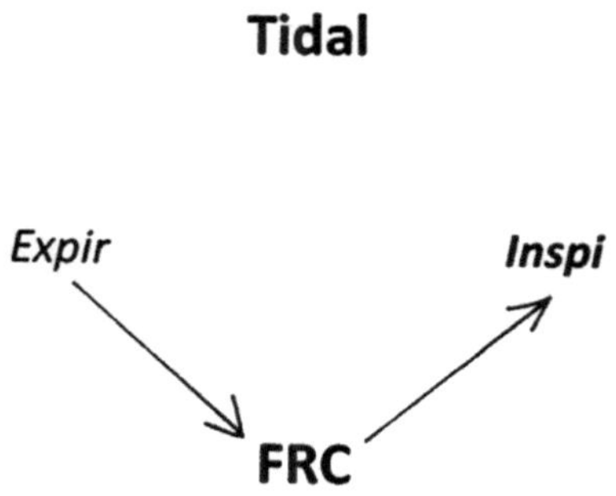

Tidal volume is the air actively breathed in from functional residual capacity (*FRC*) or passively breathed out to FRC. The depth of the tidal volume (TV) (vertical axis of inverted triangle) is determined at the respiratory "pacemaker" neuronal network in the brain stem. A deep breath is voluntarily taken by cortical impulses that bypass the autorhythmic respiratory neurons and directly synapse with spinal motor neurons that innervate respiratory muscles that include the diaphragm

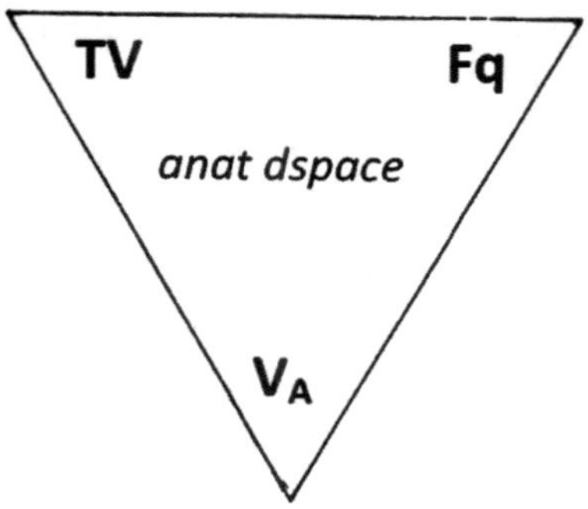

Alveolar ventilation (V_A) replenishes the alveoli and V_A is determined by the tidal volume and the frequency of breathing. In patients, compensatory effort to maintain alveolar ventilation is seen in shallow (low TV), rapid (high frequency, *Fq*) breathing or slow (low *Fq*), labored (high TV) breathing

intimately with each other, separated by a very thin pleural space. Given a chance, the lungs will tend to collapse inwards. Even at the end of a normal breath (expiration). The tendency of the lungs to recoil inwards is balanced by the equal and opposite outward recoil of the chest wall. This generates a negative Pip in the pleural space. In situations when there is physical trauma and air enters the pleural space (pneumathorax), the lungs will be held open longer by the chest wall.

When we do a voluntary force expiration, the lungs do not empty completely, and the minimum lung volume is the residual volume. At this volume, the outward recoil of the chest wall is increased and higher than at the lung volume after normal expiration (functional residual capacity, FRC). With a maximal inspired effort, the lung volume reaches the total lung capacity (TLC). At TLC, the chest wall recoil is inwards, in the same direction as the recoil of the elastic lungs. The chest wall mechanic is slightly different as it is a rigid structure and the direction of recoil can change depending on the position of the chest wall (a good illustration is the breast bone of chicken; press it inwards and it spring outwards; stretch it outwards and it will recoil inwards. Try it and think about FRC and TLC the next time you eat at a KFC!)

3. How does the alveolar pressure and Pip compare at the beginning and end of a breath?

Answer The alveolar pressure at the start and end of inspiration is the same as the atmospheric pressure. For the Pip, it becomes more negative.

Concept The student needs to think about pressures when thinking through lung function. There are four different pressure terms to use when describing how each of them changes during a respiratory cycle. The card showing four of clubs is a helpful reminder, the club resembling like the pulmonary alveoli! Physiology texts will show a graph of these pressure changes during inspiration and expiration. The four pressures are alveolar pressure, Pip, atmospheric pressure and transpulmonary pressure. The alveolar pressure should be the same as atmospheric pressure at the beginning and end of inspiration (or end and start of expiration, respectively). The Pip is always negative at rest. During a respiratory cycle, it becomes more negative during inspiration when the inspiratory muscles contract. During expiration, it returns to its negative preinspiratory value. The transpulmonary pressure is the transmural pressure (mural means wall, in this case, the alveolar wall) that is calculated. It is the expanding or distending pressure, and the value is the difference between the intraalveolar pressure and the Pip. Thus, the transpulmonary pressure is the highest at the end of normal inspiration (a tidal volume has entered the lungs) when the Pip is most negative and the alveolar pressure is 0 mmHg.

4. How is the rate of airflow related to airway radius in a *single* airway?

Answer Airflow is proportional to the airway radius to the power of 4.

Concept The aerodynamic principle in airways is the same as the hemodynamic relationship in blood vessels. Airflow (F) is a single airway which is predominantly affected by changes in the airway radius (r). The airway Resistance® is inversely related to the power of 4 of the radius. This means that if the airway radius is halved,

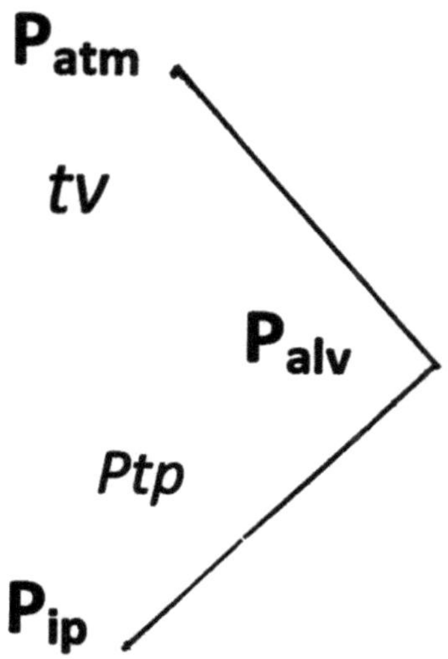

Humans inspire by negative-pressure breathing. The alveolar air pressure becomes negative (subatmospheric at ref 0 mmHg). Inspired tidal volume then enters the lungs down the atmospheric-alveolar pressure gradient. The "negative" P_{alv} is achieved when the inspiratory muscles contract and make the intrapleural pressure more negative. As such, the transpulmonary pressure increases and expands the alveoli and the P_{alv} becomes negative

the airway resistance is increased by 16 times. Since F changes proportionately with R, the airflow will then increase or decrease markedly, in parallel with the power of 4of the radius. The airway has smooth muscles. The bronchoconstriction or bronchodilation occurs predominantly in the noncartilaginous segment of the respiratory tree. An autonomic parasympathetic nerve releases acetylcholine that produces bronchoconstriction, while the sympathetic nerve action acts on beta adrenergic receptors to increase the airway radius (persons with a history of asthma commonly have their handbag/pocket an inhaler containing a beta receptor agonist). It should be noted that a large part of sympathetic bronchodilatory effects is mediated indirectly via stimulation of adrenaline secretion from the adrenal medulla by the sympathetic nerve. The circulating adrenaline acts on the same beta receptors on airways as the neurotransmitter noradrenaline released by the sympathetic adrenergic nerve to the airways.

The word single in the question is *italized* for an important reason. When we consider respiratory aerodynamics or vascular hemodynamics, the principle is applied to a single air or blood conduit. When we take the whole respiratory tree with its multiple generations of branching into smaller and smaller airways leading into the alveoli, the actual airway resistance is no longer inversely related to the radius of a particular segment of the respiratory tree. Branching of airways into parallel flows down the smaller airways, increases the total cross sectional area of the air passages, and has the effect of decreasing the downstream regional airway resistance. Thus, the highest airway resistance is not located at the deepest part of the respiratory tree with the smallest airways. Note, however, that the airflow ml air per min in each of the lungs should still be the same at every generation of the respiratory tree.

5. How is spirometry used to assess the major airway parameter that affects airflow?

Answer Spirometry assesses the airway resistance by measuring the forced expiratory volume in one second $FEV_{1.0}$.

Concept Spirometry is able to determine lung volumes and lung capacities except the residual volume (RV). Since FRC and TLC include the RV, the FRC and TLC are not determined by spirometry either. The forced expiratory volume is basically a forced vital capacity measurement. Vital capacity is obtained by asking the patient to breathe in maximally and then from TLC, to breathe out maximally to RV. For a forced VC, the expiratory phase is done by the patient as forcefully and as rapidly as possible. The volume of the vital capacity that can be expelled by the forced expiratory effort in the first second is then noted. For a healthy lung with normal airway resistance, the FEV in one second should be at least 70 % of the vital capacity (VC). For increased airway resistance, e.g., asthma, the $FEV_{1.0}$ will be less than 70 %. Pulmonary disease is clinically placed under two categories, namely obstructive and restrictive lung dysfunction. The definition of an obstructive pulmonary condition is when there is an abnormal increased airway resistance. Thus, the $FEV_{1.0}$ spirometry is commonly performed in patients with asthma and other obstruction lung disease. The restrictive lung problems are due to a characteristic reduced lung/chest wall compliance. Any pathophysiologic or anatomical situations that restrict the normal expansion of the lungs are called "restrictive."Words and definitions can sometimes be unhelpful in understanding the physiology. A student can, not incorrectly, view increased airway resistance in obstructive pulmonary disease as "Restricting" the airflow!

6. How does a *slow*, deep inspiration affect airway resistance?

Answer Airway resistance decreases with increasing lung volume.

Concept The main parameter that determines airway resistance is the airway radius. Airway resistance is inversely related to the power of 4 of the radius. One of the pressure that alters airway radius is the transmural pressure across the wall of the airways. This airway transmural pressure is the difference between the pressure in the airways and the pressure outside the airways, i.e., the Pip. The transmural pressure is a distending pressure. The higher the transmural pressure, the larger the radius will increase, and consequently the airway resistance will decrease. A deeper breath or increased tidal volume is achieved with a greater voluntary inspiratory effort. This is associated with an increased negativity of the Pip. The airway transmural pressure is then greater and the airway resistance lower. This makes physiologic sense since when we feel encouraged and feel "inspired to inspire" deeply, a larger volume of fresh air is more easily taken into the lungs! The italicized "slow" in the question also makes an aerodynamic point. An increased airflow velocity tends to increase the resistance to airflow due to turbulent airflow. So a hurried maximal inspiration will not reduce the airway resistance as much as a leisurely deep breath. In patients with obstructive lung disease, there is pathophysiologic elevated airway resistance. Characteristically, the patient's breathing pattern is slow and deeper. The slower, labored breath partly helps to avoid adding to the airflow resistance load. If you believe in the benefits of the Chinese physical exercise *tai chi*, think of the slow, relaxed, deep breathing that accompanies some of the movements!.

7. How does the transpulmonary pressure vary in an upright lung at the end of normal expiration?

Answer In an upright lung, the transpulmonary pressure at the end of a normal expiration increases progressively from the base to the apex of the lungs.

Concept The transpulmonary pressure is the transmural pressure at the alveolar wall. In the upright lung, the Pip is affected by gravitational force. The Pip is no longer the same at every level of the lungs from the base to the apex. Due to the weight of the lung tissue, the Pip at the base of the lungs is less negative than the apex of the lungs. This translates to a higher transpulmonary pressure at the apex than at the basal alveoli at the end of a normal expiration (FRC). The size of the apical alveoli are bigger than the basal alveoli. However, alveolar ventilation is not reflected in the alveolar size at FRC. What is important in ventilation is how much of the next tidal volume (fresh, inspired air) is distributed to the different regions of the upright lung. The larger apical alveoli are surprisingly less compliant than the basal alveoli (In physiology text, the apical alveoli are shown as located on the upper plateau portion of the lung compliance curve, whereas the basal alveoli are sited at the more compliant steep slope of the curve.). Thus, more of the tidal volume actually is distributed to the more dependent part of the lungs, i.e., these alveoli are more ventilated or replenished with new, inspired air. What does all this mean in terms of respiratory function? Is this just a trivial detail of no consequence? The different degree of alveolar-ventilated parts of the normal, upright lungs will affect the oxygenation, since the pulmonary blood flow in the upright lung is also affected by gravity (oxygenation is dependent on the ventilation/perfusion or V/Q matching). I tell my students after they have understood how alveolar ventilation is better at the base of the upright lung, "things are not always better at the top!"

8. What are the two major factors that contribute to lung recoil?

Answer Lung elasticity and alveolar surface tension.

Concept The work of breathing, when airflows require energy. The energy is expended as airflow against airway resistance and against the opposing elastic force to alveolar expansion. One major normal resistance factor to lung expansion is the lung recoil force. The lungs are expanded from birth *ex utero* and the lungs always have an inward recoil tendency to collapse into its natural deflated state. Lung recoil is due to its elasticity. Any physiological structure that is elastic has an elastic recoil. Thus, the elastic recoil opposes distention or stretch (this use of the word "elastic" is different from common usage that means an elastic item, e.g., pajamas are easily stretchable!). The second component of lung recoil is the surface tension force of the alveoli. The surface tension (ST) force is due to the thin layer of fluid that lines all alveoli. The water–air interface accounts for the ST force. So when the alveoli expands, they expand against the elastic recoil and the recoil surface tension force. The essential pulmonary surfactant has a ST-lowering action and produces the alveoli stability that prevents ST force from collapsing in particular, the smaller alveoli at the end of normal expiration. The relationship between alveolar ST, the radius of the alveolus and the distending pressure needed to keep it expanded is given by Laplace's law. In premature babies, the pulmonary surfactant is deficient and the lung recoil of the baby's lungs is high due to the greater ST force. The babies suffer from breathing difficulty, described as "respiratory distress syndrome."

9. Which two *changing* parameters determine the alveolar ventilation?

Answwer Tidal volume and frequency of breathing

Concept Not all the volume of each inspired air (tidal volume) reaches the alveoli where respiratory gases are exchanged. Part of the respiratory tree serves as conducting conduits for the airflow besides protective and immune defence functions (the lungs are in direct exposure to any harmful environmental stimuli. This nonrespiratory space is called "anatomical dead space" (ADS, the "dead" is an unfortunate term for a living functional part of the lungs). To calculate ventilation of the respiratory zone ("alveolar ventilation"), the tidal volume minus the ADS is multiplied by the frequency of breathing. The ADS is roughly equal to the person's body weight in pounds. Compared to the changing values of tidal volume and respiratory rate, the ADS is relatively unchanged. For patients with poor tidal volume (restrictive lung disease), since the ADS is fixed, a reduced volume of each breath reaches the alveoli exchange area. Thus, characteristically, the patient has compensatory increased breathing rate. The breathing is described as shallow and rapid. For those who enjoy snorkeling and admiring created beauty under the sea, the breathing tube represents an extension of the ADS in the snorkeler! The depth of submersion underwater during snorkeling is, thus, limited by the need for adequate alveolar ventilation. Those who are stimulated watching medical dramas will hear about the emergency surgical procedure called "tracheostomy". For persons who developed acute respiratory failure, the tidal volume drops markedly. The tracheostomy surgically reduces the ADS and aids to sustain the airflow and oxygenation at the alveoli.

Humpty Dumpty sat on Alveolar Wall

And he observed how the PO_2 fall

Input Venti 160 , Output Blood Oxyg 100,

Alveolar air PO_2 is replenished again and again.

Not all the air in a tidal volume reaches the alveolar respiratory zone. The alveolar air PO_2 is an equilibrium between fresh, high PO_2 inspired air and the diffusion of oxygen into pulmonary alveolar capillary blood

10. What accounts for the abnormal lung compliance in emphysema?

Answer Destructive loss of elastic tissues.

Concept Emphysema is an obstructive lung disease that is accompanied by destruction of lung tissues. The loss of lung tissue included elastic structures. This leads to less elastic recoil of the emphysematous lung. Lung compliance is defined as the unit change in lung volume per unit change in transpulmonary pressure. Compliance is thus distensibility. Since elastic recoil of the lung opposes lung expansion, lung compliance is inversely related to its elastic recoil. The emphysematous lung, therefore, has an abnormal increased compliance (the pulmonary compliance curve is shifted to the left). In emphysema, supporting tissues that help to keep the airway patent are also destroyed. This results in the tendency for the airway to collapse, especially during expiration and increasing the airway resistance. With chronic emphysema, the phenomenon of "airtrapping" occurs with the markedly elevated airway resistance during expiration. The patient will have an enlarged FRC. The distended FRC is also caused by the abnormal lung compliance in emphysema which accommodates the "air trapping."An important strong contributing factor to the development of emphysema is smoking. This detrimental lifestyle habit needs to be "emphasized" in health and wellness education!

Chapter 7
Upright Lung, Ventilation, and Blood Flow

Lung function involves respiratory airflow and also pulmonary blood flow. Moreover, the airflow and blood perfusion are interfaced at the alveolar-capillary membrane, where respiratory gases, oxygen, and carbon dioxide are exchanged. The diffusion of O_2 and CO_2 at the lung is thus influenced not only by alveolar ventilation but also by a matched pulmonary blood supply to the alveoli. The lungs are relatively large structures and the body posture modifies both the ventilation and blood perfusion at different levels above the heart in an upright lung of a standing person due to gravitational force.

1. How does the ventilation/perfusion (V/Q) ratio change in an upright lung?

Answer The ventilation perfusion ratio increases from the base to the apex of the upright lung.

Concept The alveolar ventilation is best at the base of the upright lung and decreases towards the apex. The pulmonary blood hydrostatic pressure is also affected by gravity. Thus, pulmonary blood flow or perfusion is also best to the basal alveoli and lessen towards the apical alveoli. The gradient in the progressive reduction in pulmonary perfusion is steeper than for the alveolar ventilation. Thus, the calculated V/Q matching ratio increases from the base to the apex of the upright lungs. For the whole lungs, the V/Q balance has a value ~0.8, where alveolar ventilation is 4 L/min and pulmonary blood flow is 5 L/min, the ejected blood flow rate (cardiac output) from the right ventricle. So, there is uneven regional V/Q in the resting lungs. When a person does physical activity, the pulmonary blood flow is more evenly distributed throughout the lungs and the V/Q ratio is less uneven across the lungs. Why does the V/Q ratio matter? We can imagine the two extremes of V/Q mismatching—zero ventilation but normal perfusion which will give a V/Q of zero, or absence of perfusion but normal ventilation which has a V/Q value of infinity. These two scenarios provide no oxygenation of the pulmonary blood. Thus, the optimal crosstalk between the airflow and blood flow is essential for effective exchange of oxygen as well as for the adequate removal of carbon dioxide from deoxygenated mixed venous blood from the periphery.

H. M. Cheng, *Physiology Question-Based Learning*, DOI 10.1007/978-3-319-12790-3_7

2. How does the pulmonary blood flow change if a voluntary deep inspiration is done?

Answer The pulmonary blood flow should physio-*logically* increase.

Concepts In Chap. 1, we touched on the way in which increased lung volume is associated with a decreased airway resistance which physio-logically aids airflow during a deep inspiration. The *net* effect of a deep inspiration is also to increase pulmonary blood flow. The word "net" indicates that there are multiple factors impinging on pulmonary perfusion during a maximal inspiration. A positive effect is due to the decrease in the central venous pressure when the thoracic space is more expanded. The venous return into the right ventricle is thus enhanced and this provides the greater cardiac output/pulmonary blood flow to the lungs. The effect on pulmonary vascular resistance (*pvr*) is a little unique and unusual. The lowest *pvr* is at functional residual capacity (FRC), at the end of a normal expiration. On either side of FRC, of lung volume decrease or increase, the *pvr* increases. In some physiology texts, this is explained by the differential changes in vascular resistance in alveolar vessels and extraalveolar vessels. The alveolar vessels are directly subjected to the mechanical compression of the alveolar air pressure while the radius of the extraalveolar vessels changes with changes in the intrathoracic pressure. Thus, during a deep inspiration, the extraalveolar vascular resistance lessens while the alveolar vessels are more compressed by the larger alveoli. Going back to the answer and "net" effect above, overall, there is an increase in pulmonary blood flow with a deep inspiration. This is also called a "respiratory pump" effect and contributes to the greater cardiac output during exercise.

For the physiology teacher, the issue is "should we teach the phenomenon of the graphically represented U-shape *pvr* changes during a respiratory cycle?" Is telling our students about the "respiratory pump" effect good enough? Depending on the structure of each medical school curriculum and the available time allocated for small group discussion/tutorials, the teacher will have to decide discerningly. Do we teach more to educate, to inspire curiosity and discovery, or merely impress our students?!

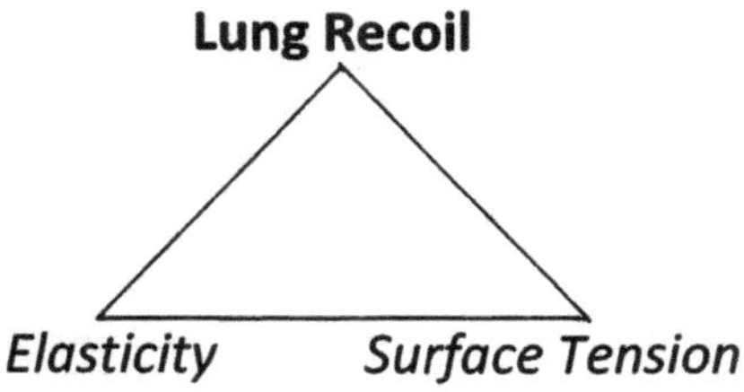

The tendency for the lung to recoil is due to two factors. The elasticity of the lung tissues produces the elastic recoil. Note that elasticity is defined physiologically as the recoil force that opposes distention or stretch. Lung elasticity is inversely related to lung compliance. The second factor is the alveolar surface tension, which is an inward collapsing alveolar force

3. What is the physiological benefit of the response of the pulmonary blood vessels to local hypoxia?

Answer Hypoxic pulmonary vasoconstriction helps to compensate for V/Q mismatching.

Concept The pulmonary vasculature is unique in its response to hypoxia. All other arterioles in the body, skeletal, coronary, cerebral will vasodilate in response to hypoxia. This aims to ensure that the muscle, heart, or brain receives adequate oxygen. In contrast, pulmonary vessels constrict when they sense a decrease in oxygen. The bronchial circulation supplies oxygenated blood for the metabolic needs of the lungs. The pulmonary circulation, however, recycles blood from the periphery to be oxygenated at the alveolar-capillary interface. In response to the local hypoxia, the pulmonary vessels constrict, and this has the effect of redirecting more of the blood to the alveoli that have normal PO_2. Thus, if the regional alveoli are relatively underventilated (V/Q is reduced), the alveolar PO_2 will be less than 100 mmHg, and compensatory hypoxic pulmonary vasoconstriction (HPV) occurs.

In a different situation when the V/Q is high (relatively overventilated alveoli as when the alveoli receive a reduction in blood flow), the partial pressure of carbon dioxide in alveolar air is decreased. The local hypocapnia will produce bronchoconstriction of the corresponding local airways and airflow is channeled to alveoli that are better perfused.

HPV accounts for pulmonary hypertension when there is generalized vasoconstriction in response to hypoxia throughout the lungs. This is seen in persons exposed to high-altitude hypoxia.

4. State reasons whether you expect the pulmonary arterial pressure to increase much during exercise.

Answer The pulmonary arterial pressure P_{pulm} does not increase much as the *pvr* is decreased by any increase in P_{pulm}.

Concept The pulmonary hemodynamics is different from that of the systemic circulation. In systemic circulation, we have arterial blood pressure = cardiac output × total peripheral resistance. Changes in the two determinants, cardiac output (CO) or total peripheral resistance (TPR) will change the arterial blood pressure. The equation is mechanistically explained generally from a left to right direction. In the lungs, the pulmonary blood vessels are relatively compliant and easily stretched by increasing the blood pressure. When the right heart pumps a greater cardiac output, the pulmonary arterial blood flow is increased. We might expect a proportionate, parallel increase in the pulmonary arterial pressure. However, the increased volume of blood flow mechanically distends the pulmonary vessels. In addition, some pulmonary vessels that are not fully patent at rest are recruited to supply the increased pulmonary blood flow. These two "mechanical vasodilation" events, vascular distention and vessel recruitment lowers the pulmonary arterial pressure. Thus, the pulmonary hemodynamic can be said to proceed in a left to right direction when we look at the equation, pressure = flow × resistance; however, with the difference that any potential increased blood pressure will decrease and not increase the vascular resistance.

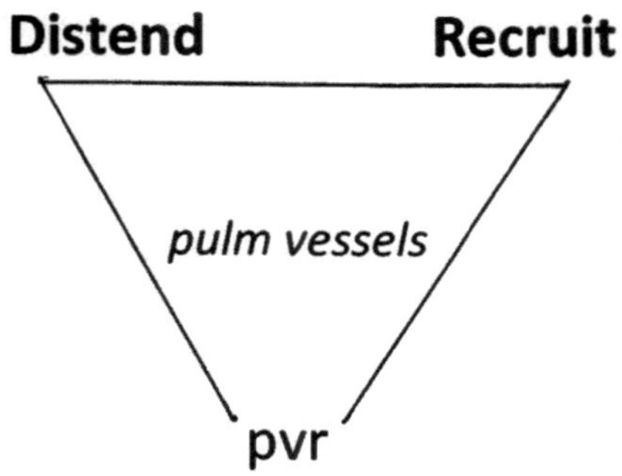

The pulmonary blood vessels are relatively compliant. As such, the pulmonary vascular resistance (*pvr*) is decreased when the blood pressure in the pulmonary artery increases. Pressure or mechanical distention and recruitment of more blood vessels vasodilate, and the pvr to blood flow is reduced

5. How does the pulmonary blood flow contribute to increased oxygenation (ml O_2/min) during increased metabolic demand?

Answer Increased pulmonary blood flow overcomes the perfusion-limited oxygenation in the lungs during rest.

Concept This question might appear to ask the obvious. A number of points need to be made. First, oxygenation is different from ventilation. Oxygenation is ml oxygen/min, whereas ventilation is simply ml air/min. Oxygenation occurs rapidly across the alveolar-capillary membrane, down the PO_2 gradient. Alveolar air PO_2 is 103 mmHg and mixed venous blood has PO_2 of 40 mmHg. Complete oxygenation at rest by net diffusion and equilibration of PO_2 is achieved at less than half the pulmonary capillary transit time (time of exposure to the alveolar air). No further diffusion and oxygenation of blood takes place beyond the equilibration point. More oxygenation (ml O_2 per unit time) can occur if there is more inflow of deoxygenated blood (perfusion-limited oxygenation). An illustration to help appreciate this is as follows: Each unit of blood represents an airplane and the alveolus is where the air passengers get on. The plane has a limited seating capacity (103 passengers). In any given time, if there is a need to fly more passengers to the periphery of the cardiovascular world, you simply increase the number of your airplanes at the alveolar air terminal, since each plane is filled up quickly by 103 passengers.

This also makes the point that during exercise, the main factor that provides increased oxygen delivery to the cells is not hyperventilation but increased pulmonary blood flow. Hyperventilation will only increase the alveolar air PO_2 slightly and this does not result in a marked increase in saturation of hemoglobin and the blood oxygen content (sigmoid-shape of oxygen–hemoglobin dissociation curve, see Chap. 3).

6. How does the PO_2 gradient component of diffusion capacity of oxygen in the lungs (D_LO_2) increase during exercise?

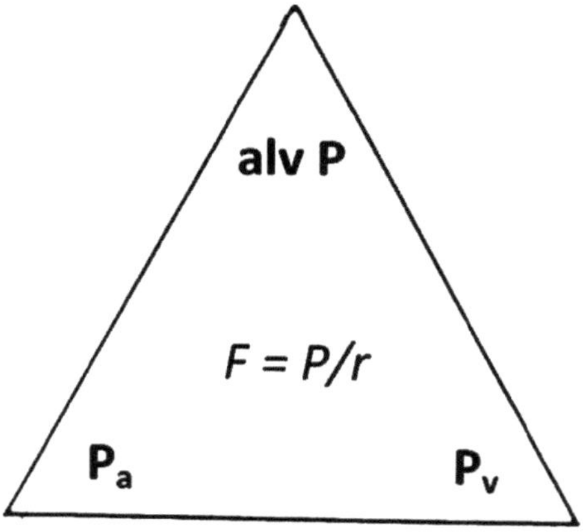

The pulmonary blood flow is determined by the driving perfusion pressure, which is determined by different parameters at different levels of the upright lungs. At the basal lung region, P_a—P_v drives the flow. In the mid-region of the upright lungs, P_a—alveolar air pressure (*alv P*) determines the flow. At the apex, P_a has dropped to a situation when alv P—P_a becomes the deciding perfusion pressure

Answer The PO_2 gradient is increased mainly due to the reduction in the mixed venous blood PO_2 due to increased metabolism.

Concept Lung oxygenation is described by the index D_LO_2 which is the volume of oxygen/min transferred across the alveolar-capillary membrane per mmHg (mmHg here is the difference between alveolar PO_2 and *mean* pulmonary capillary PO_2; *mean* because, there is no one single value as the blood PO_2 rapidly equilibrates with that in alveolar air). The student might think that during exercise, with hyperventilation, the alveolar air PO_2 should be raised above 103 mmHg. However, the arterial blood and thus the alveolar air PO_2 is relatively constant during physical activity. The gradient for oxygen diffusion is steeper due to the arrival of venous blood that is more deoxygenated (PO_2 less than 40 mmHg).

The same situation also applies to the carbon dioxide diffusion from the pulmonary blood into the alveolar air. The alveolar air PCO_2 is still about 40 mmHg. The increased production of metabolic CO_2 increased the mixed venous PCO_2 to be higher than 46 mmHg, the value at rest. Note the starting partial pressure gradient for CO_2 diffusion is much less (6 mmHg) than that for O_2 (60 mmHg). This is due to the higher solubility of CO_2 in blood.

The diffusion of both respiratory gases is a passive process and as described by Fick's law of diffusion—surface area is the other major determinant for the passive movement of O_2 and CO_2. The surface area available for diffusion at the alveolar-capillary membrane comprises the apposition of the alveolar epithelium and the endothelial cell of the pulmonary capillary, separated by a thin interstitial space.

7. What enzymatic action associated with the pulmonary circulation plays a role in blood pressure control?

Answer Conversion of angiotensin I (AI) to angotensin II (AII).

Concept Besides gas exchange, the lungs have a number of nonrespiratory functions. One essential reaction occurs at the endothelial cells of the pulmonary blood vessels. This is the major location of the enzyme, angiotensin-converting enzyme

(ACE). The ACE transforms AI to the bioactive AII. AII is a multitasker peptide hormone. AII is a potent vasoconstrictor, and this increases the TPR. In the blood pressure formula (BP = CO × TPR), the CO determinant is affected by blood volume. AII hormone indirectly has a crucial role in maintaining the blood volume. This is via the control of sodium balance by its action on stimulating the secretion of the adrenal cortical hormone aldosterone (note the adrenal medullary catecholamines that have effects on blood pressure are stimulated by sympathetic nerves). Sodium balance is the main determinant of extracellular fluid (ECF)/blood volume. In addition, the AII has been shown to be a dipsogenic (increases thirst/water intake) and stimulated vasopressin release from the posterior pituitary. The use of ACE inhibitors in prescription for hypertension is well known. The kidneys are the source of the enzyme/hormone renin that cleaves a plasma protein, angiotensinogen to generate the AII precursor AI. AII itself has a direct effect in increasing the sodium reabsorption from the proximal tubule of the nephron. By its constricting action on the renal arterioles, the AII also conserves sodium by reducing the glomerular filtration rate (GFR) and thus the filtered sodium load. So, we have here an integrated picture of the lungs and the kidneys collaborating with the heart/blood vessels to regulate blood pressure.

It makes physiologic sense, that the lungs have a role in ensuring a normal arterial blood pressure. Oxygenated blood must circulate adequately to the tissues to be useful. Sluggish blood flow, e.g., during cardiac failure, results in a stagnant hypoxia. The driving pressure for tissue perfusion is the mean arterial pressure.

8. How does an alveolar dead space and anatomical dead space compare in terms of its CO_2?

Answer All "dead" spaces in pulmonary physiology are sites of nonexchange. Therefore, the air in "dead" spaces, whether anatomical or alveolar will resemble inspired air and the partial pressure of CO_2 will be close to 0 mmHg.

Concept Students are often "dead" confused with the terminology of some respiratory terms and one in particular is "dead" space. It is helpful to think of "dead" space as a location, where there are no exchange respiratory gases. This means that the partial pressures of gases in the inspired air at these places remain unaltered.

The reason for the nonexchange in anatomical and alveolar "dead" spaces are slightly different. All normal, living persons have anatomical "dead" space (ADS)! This is the space or anatomical volume in the airways of the respiratory tree that functions as conduits for airflow to the respiratory exchange alveoli region. Each tidal volume during expiration, thus, comprises a mixture of alveolar air and unexchanged air in the ADS. This explains why expired PCO_2 is lower than alveolar air PCO_2 (<40 mmHg). It also accounts for the fact that expired air PO_2 is higher than alveolar air PO_2 (>103 mmHg).

An alveolar "dead" space is not a normal situation. All alveoli should be involved in gas exchange. If a local region of the lungs is deprived of blood flow (e.g., pulmonary embolism), then the air of the ventilated alveoli in that nonperfused region will remain unchanged. In terms of ventilation/perfusion ratio, the V/Q has a ratio of infinity. I help students to remember the "dead" space V/Q by linking "infinity" with outer ("dead") space!

9. How might you expect the FRC and the total lung capacity (TLC) to change when standing up from lying down?

Answer The TLC remains the same while the FRC is less in the supine compared to the upright position.

Concept The TLC is the lung volume that is achieved with a maximal inspiratory effort. This maximally filled lung volume should be the same in a normal person, lying or standing up. The FRC volume is that at the end of a normal expiration. This volume is determined by two opposite recoil forces, namely the inward recoil of the lungs and the outward recoil of the chest wall. The chest wall and diaphragm is functionally considered as a single mechanical unit. When lying down, there is some compression of the diaphragm by the abdominal contents (more when a person is fat!). This means that the outward recoil of the diaphragm/chest wall is decreased. The balance of the lung/chest wall recoil will now shift to a decreased FRC. Since the TLC is unchanged in either posture, the inspiratory capacity is actually more in the supine position. This may sound unusual to the student since the word "capacity" by instinct seems naturally to be better in the standing person.

Clinically, an enlarged abnormal FRC is seen is emphysema. This obstructive lung disease is associated with a loss of elastic lung tissues. The inward elastic lung recoil is thus reduced. The balance of lung/chest wall recoil is pushed outwards to a bigger FRC. The phenomenon of "air-trapping" due to the increased airway resistance in emphysema also contributes to the larger FRC.

10. In an asthmatic attack, why does sitting up from lying in bed help to relieve the breathing difficulty?

Regional, less than optimal ventilation or blood perfusion is compensated by a corresponding reduction in the paired perfusion or ventilation respectively. This is to match the ventilation/perfusion balance, to lessen "wasted", "overperfused" alveolar blood flow or "overventilated" alveoli

Answer The lung compliance is higher in the upright lung since the alveoli are expanding against a significant larger pulmonary blood volume in the supine position (dense mesh of capillary network at each alveolus)

Concept In a normal person, the slight difference in lung compliance between lying and standing is of little consequence (all students find the horizontal position comfortable!). During labored breathing in asthmatics due to increased airway resistance (hyperresponsive airways in asthma produce the bronchoconstriction), any change in body position that helps ease the dyspnea will be assumed by the person. There is a significant more pulmonary blood volume in the supine position compared to the upright lungs as the pulmonary vasculature being relatively compliant. When sitting up, the asthmatic lungs have an improved compliance as the alveoli are expanded against a decreased "sheath" of pulmonary blood that surrounds the alveoli.

In the normal lungs, the lung compliance changes with the lung volume. Lung compliance is less at very small volumes near the residual volume. Lung compliance is also decreased near the maximal total lung capacity. Thus, the lung compliance curve (lung volume versus transpulmonary pressure) is a sigmoid with lower compliance at low and high lung volumes. A useful demonstration of this volume/compliance relationship is the action of blowing a balloon. It takes more effort to blow up the balloon at the beginning. The midway expansion of the balloon is easier, and as the balloon reaches near its maximal size, increased blowing pressure is again needed.

Chapter 8
Oxygen Respiratory Physiology

It is good to distinguish between the term “ventilation” and “oxygenation.” If one notes the different units for ventilation (ml Air/min) and oxygenation (ml O_2/min), it becomes obvious that they have quite different meanings. Airflow is dependent on the difference in air pressure. Oxygenation, however, is by simple diffusion and depends on the availability of a partial pressure gradient for oxygen (PO_2). In the liquid blood medium, the PO_2 refers only to the dissolved portion of oxygen in plasma/cytosol.

1. How is the change present in the fractional concentration of oxygen with increasing altitude?

Answer There is no change in the fractional concentration, but there is a decreasing PO_2 with increasing altitude.

Concept The partial pressure of oxygen is calculated by multiplying the fractional concentration with the atmospheric pressure. At sea level, the oxygen comprises 21 % of air composition. Thus, the partial pressure of oxygen is 0.21×760 mmHg which gives ~ 160 mmHg.

The atmospheric pressure decreases with ascent to higher altitude, but the fractional concentration of oxygen is unchanged. At the atmospheric pressure of 500 mmHg, the partial pressure will drop to 0.21×500 or 105 mmHg.

There is a gradient of PO_2 from dry inspired air to the alveolar air, from 160 mmHg to 103 mmHg. If the starting inspired air has a reduced PO_2 of 105 mmHg, the alveolar air PO_2 will decrease in parallel. This hypoxia with a low arterial blood PO_2 (hypoxic hypoxia) is due to the fall in atmospheric air PO_2.

At sea level, the theoretical maximum the PO_2 in alveolar air can reach is ~ 160 mmHg with voluntary hyperventilation. With normal breathing at rest, the arterial blood PO_2 at 100 mmHg has a hemoglobin (Hb)-O_2 saturation of ~98 %. Thus, a voluntary effort in breathing by increasing the alveolar air PO_2 will not dramatically increase the total blood oxygen content.

A different situation, where the alveolar air PO_2 might become 160 mmHg, is a nonperfused alveolus. Since oxygen is not extracted with the absence of blood

H. M. Cheng, *Physiology Question-Based Learning*, DOI 10.1007/978-3-319-12790-3_8

The alveolar PO_2 is an equilibrium value between fresh input of tidal volume air and extraction of O_2 into pulmonary capillary blood. The expired tidal volume (TV) actually has a higher PO_2 than alveolar air since the expired TV is a mixture of alveolar air and unaltered, inspired air in the anatomical dead space

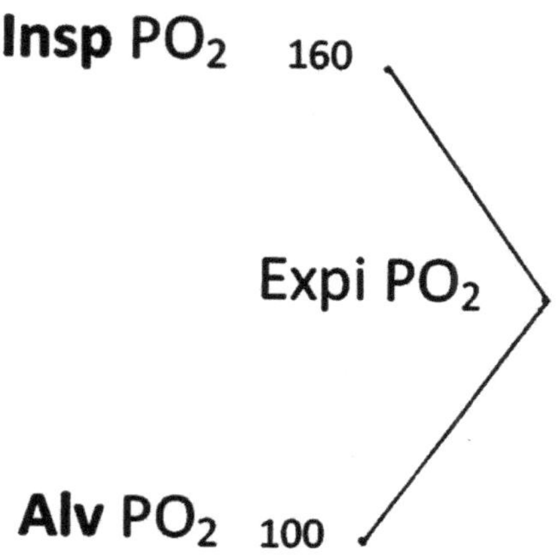

supply to the alveolus, the alveolar air PO_2 is unchanged at ~160 mmHg. Of course, the high PO_2 is the wasted ventilation to the nonperfused alveolus.

2. How is the value of partial pressure of oxygen in anatomical dead space calculated at normal body temperature?

Answer The partial pressure is calculated by multiplying the fractional concentration of oxygen with the pressure (atmospheric minus the water vapor pressure at 37 °C)

Concept Inspired air is very soon moistured within the respiratory tree. The water vapour pressure is dependent on the temperature. At 37 °C, it is 47 mmHg. Thus, the partial pressure of oxygen in the warmed, moisture air will be 0.12 (760–47) or ~150 mmHg. The anatomical dead space contains air that will not be altered as there is no exchange of the respiratory gases.

In the alveolar gas equation, the inspired PO_2 is used to calculate the alveolar air PO_2 in this equation: Alveolar PO_2 = Inspired PO_2 _ alveolar PCO_2/R, where R is the respiratory exchange ratio or respiratory quotient (rate of CO_2 production/rate of oxygen consumption)

Normally $R = 0.8$ and putting in values, Alveolar PO_2 = 150 mmHg (see Q1) minus 40/0.8, giving alveolar PO_2 as 100 mmHg.

Graphically in physiology texts, the relationship between alveolar air PCO_2 and PO_2 described by the alveolar gas equation is shown as the CO_2 (y-axis)–O_2 (x-axis) diagram. The relationship is an inverse one ranging from a value similar to mixed venous blood (PCO_2 46, PO_2 40 mmHg in a nonventilated alveolus) to inspired air (PCO_2 0, PO_2 100 mmHg, in a nonperfused alveolus).

3. How does the PO_2 change in its value along the pulmonary capillary blood?

Answer The PO_2 in the pulmonary capillary exposed to alveolar air rises from 40 mmHg to 103 mmHg.

Concept Deoxygenated blood in pulmonary arterial blood is pumped by the right ventricles to the lungs. This mixed venous blood has a PO_2 of 40 mmHg and the hemoglobin is 75 % saturated with oxygen. At the alveolar capillaries, rapid equilibration takes place and the blood PO_2 increases to reach 103 mmHg, the PO_2 in

The diffusion capacity of oxygen in the lungs ($D_L O_2$) across the alveolar-capillary (a/c) membrane is the rate of oxygen diffusion per mmHg O_2 partial pressure difference across the membrane. Note that the pulmonary alveolar capillary PO_2 progressively increases along the capillary. Total available surface area of the a/c membrane is also a major factor in passive O_2 diffusion

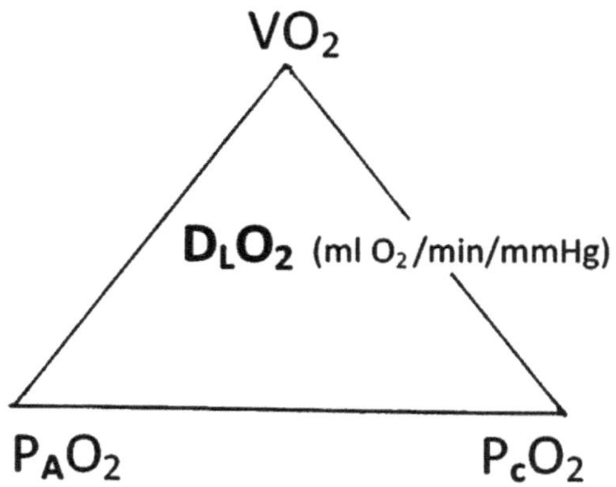

alveolar air. This net diffusion of oxygen down the PO_2 gradient across the alveolar capillary membrane is completed in half of the alveolar capillary blood transit time.

In the diffusion capacity of oxygen in the lungs ($D_L O_2 \sim VO_2/PO_2$ gradient), the partial pressure denominator is the difference between alveolar air PO_2 and the *Mean* alveolar capillary PO_2. The *Mean* value is used because the PO_2 gradient progressively decreases along the alveolar capillary.

The value for alveolar PO_2 at 103 mmHg in reality also represents an average alveolus in the whole lungs. When we consider the upright lungs and the normal ventilation/perfusion difference, the actual value of the alveolar air PO_2 is more than 103 mmHg in the apex and less than 103 mmHg at the basal alveoli (note this is not the same as alveolar ventilation being better at the basal alveoli and oxygenation also better at the base of the lungs).

The blood that leaves the lungs in the pulmonary vein represents the sum of blood from all regional alveoli in the upright lungs and has an average arterial blood PO_2 of 100 mmHg (and arterial PCO_2 of 40 mmHg). Thus, the alveolar air PO_2 of 103 mmHg might be viewed as in a representative alveolus in the supine person when gravity does not produce the varied ventilation/perfusion (V/Q) distribution across the upright lungs. PO_2 of 103 mmHg is that of air in the resting "horizontal" alveoli.

4. What does "perfusion-limited" diffusion of oxygen mean at the alveolar capillary membrane?

Answer Since rapid equilibration of PO_2 occurs at the alveolar capillary, further oxygenation is possible only if the flow of deoxgenated blood into the lungs increases.

Concept Diffusion of respiratory gases occurs by simple diffusion, and net diffusion continues down the partial pressure gradient until the partial pressure equalizes. In normal lungs, the exposure time of deoxygenated blood to the alveolar air is less than one second. Rapid equilibration of PO_2 is achieved in less than 0.5 s. Beyond the point of PO_2 equilibrium at 103 mmHg, there is no further diffusion and thus no further oxygenation.

In pulmonary disease with "alveolar-capillary block," the diffusion distance is increased due to thickening of the alveolar capillary membrane. If the reduction of

The total oxygen content in blood is the sum of the dissolved O_2 and the hemoglobin-bound O_2. The dissolved O_2 is determined by the partial pressure of O_2. The dissolved O_2 is in equilibrium with and determines the amount or saturation of hemoglobin-bound O_2

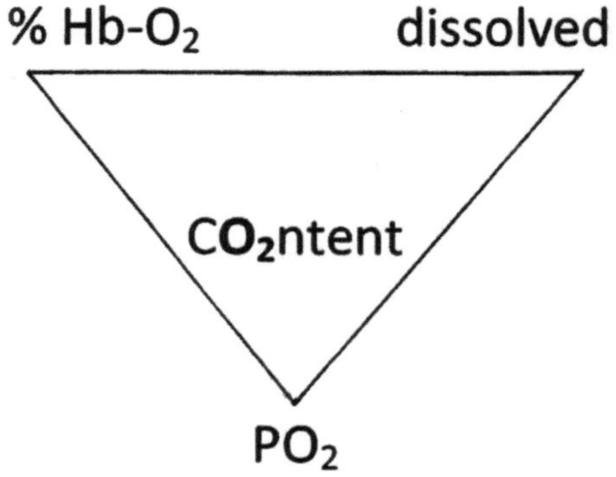

diffusion is severe, equilibration of PO_2 is not reached. The pulmonary capillary blood PO_2 remains less than 103 mmHg; the lower the value, the greater the restriction in oxygen diffusion. This situation is also called "diffusion limited."

The perfusion-limited oxygenation in the lungs at rest is overcome during exercise with a greater pulmonary blood flow when the right ventricle supplies a bigger cardiac output.

In a similar way, the unloading of oxygen at the tissues can also be said to be "perfusion limited." At the cells, oxygen is extracted from arterial blood in the capillary as oxygen diffuses rapidly down its partial pressure gradient (100/40 mmHg). When the capillary blood near the venous end is reduced to 40 mmHg, no further unloading of O_2 occurs if we take the tissue PO_2 at 40 mmHg. In order for more delivery and release of oxygen from blood to the cells, a greater flow of oxygenated blood into the microcirculation must occur.

The increased skeletal muscle blood flow during exercise fulfills this function. This "perfusion-limited" diffusion of oxygen at the tissues also explains why in sluggish tissue blood flow (e.g., in cardiac failure or blocked regional blood vessel), the cells do not extract more oxygen. In fact, the cells are deprived of sufficient oxygen is what is known as stagnant hypoxia.

5. What accounts for the normal difference in PO_2 in alveolar air and arterial blood?

Answer The physiological shunt for oxygen is due to a little mixing of deoxygenated blood from the cardiac veins and bronchial circulation.

Concept The alveolar capillary blood equilibrates with the oxygen rich alveolar air at PO_2 of 103 mmHg. The value of PO_2 in arterial blood is normally slightly lower at around 97 mmHg. This normal decrease from alveolar air to arterial blood is called physiological shunt. The accepted difference in this A-a PO_2 should not be greater than 15 mmHg.

The reason for the physiologic shunt is due to anatomical reasons. Some of the bronchial blood flow does not get reoxygenated and mixes with oxygen-rich blood draining from the lungs. In addition, some coronary venous blood also enters directly into the left ventricles.

If there is an anatomical defect producing, in this case an abnormal anatomical shunt, then the alveolar-arterial PO_2 will be increased. This occurs, e.g., in an atrial septal defect resulting in what is called a right to left shunt. A portion of the dexyogenated blood in the right atria is shunted into the left atria and dilutes the PO_2 of the oxygenated blood that is pumped out via the aorta and arteries.

Shunting of blood can also happen within the lungs in intrapulmonary shunt. The alveoli at intrapulmonary shunt are not well ventilated (the ventilation/perfusion ratio will be reduced). The pulmonary blood draining these underventilated alveoli (e.g., when a branch of the respiratory tract is blocked) will have decreased PO_2. This underoxygenated blood will then mix with blood from normal ventilated alveoli. The arterial blood PO_2 will thus be depressed, and the A-a PO_2 difference widens.

6. State the axes in the oxy-hemoglobin dissociation graph.

Answer The *x*-axis is the partial pressure of oxygen in mmHg and the *y*-axis is the hemoglobin saturation in percentage.

Concept The *x*-axis is the variable or we could say the "cause" and the *y*-axis is the effect of changing the x values. Relationships are expressed and specified by *x*-y values. In the oxy-hemoglobin dissociation graph (not wrong to call it association graph actually!), the key point is that the percentage Hb-O_2 saturation is dependent on the PO_2. Increasing PO_2 will give a higher HB-O_2 saturation.

The PO_2 is the dissolved oxygen and this dissolved O_2 is in equilibrium with the hemoglobin-bound O_2. It is important to understand that the PO_2 determines the Hb-O_2 saturation, but the reverse is not correct. The amount of oxygen bound to Hb does not affect the PO_2. Note that at arterial blood PO_2 of ~100 mmHg, the HB-O_2 saturation is already ~98 %. At venous and tissue PO_2 of 40 mmHg, the Hb-O_2 saturation does not drop to 40 % but is still at 75 %. This indicates that there is still a large reserve oxygen content in venous blood.

The Hb-O_2 curve is not a linear line but shows a sigmoid shape due to the unique cooperative binding reaction between hemoglobin and O_2. One cellular benefit of this is that at the PO_2 of 60 mmHg (a big decrease of 40 mmHg from normal arterial PO_2), the Hb-O_2 saturation is still at ~90 %. This providential coincidence allows mountain climbers to enjoy their high altitude adventures.

The student should note that the steep portion of the Hb-O_2 curve is around PO_2 of 40 mmHg. This is the operating PO_2 region at the tissues and thus an advantageous characteristic of the Hb-O_2 interactions. At the tissues, small changes in PO_2 will produce a greater change in the Hb-O_2 saturation. In other words, the unloading of oxygen to the cells, if the PO_2 decreases a small amount to 38 mmHg, is significant.

Blood oxygen bound to hemoglobin is around 98 % of the total blood O_2 content, with a small percentage dissolved in blood. Therefore, if the *y*-axis is changed to the blood oxygen content (ml O_2/dl blood), the shape of the Hb-O_2 graph will remain identical. In iron-deficiency anemia, the total blood oxygen content will be reduced. But, since the hemoglobin is normal, the Hb-O_2 saturation at any given PO_2 will not be altered in this type of anemia.

Playing cards can be used to review and enjoy respiratoy physiology. There are several numbers in respiratory function to note. Here, the students will recognize that the PO_2 103–PO_2 97 mmHg represent the normal, physiologic shunt which is due to normal anatomical architecture. A little deoxygenated blood in certain bronchial and cardiac blood vessels minimally dilutes the oxygenation in the systemic arterial blood

7. State Bohr's effect in the lungs.

Answer In the lungs, the lower partial pressure of CO_2 increases the hemoglobin affinity for oxygen and enhances the uptake of oxygen from alveolar air into pulmonary capillary blood.

Concept The Bohr's effect explains how the partial pressure of CO_2 affects the blood oxygen content. A higher PCO_2 (>40 mmHg) decreases the Hb-O_2 affinity. This occurs in the tissues where the local PCO_2 is 46 mmHg. More unloading of oxygen from the arterial blood is promoted and the blood oxygen content is reduced.

Conversely, when the blood is reoxygenated in the lungs, the PCO_2 returns to 40 mmHg. The Hb-O_2 affinity becomes higher again to ease the oxygenation of blood in the lungs. Bohr's effect is also accounted for by changes in pH. The metabolically active cells have a local pH of less than 7.4. A decrease in pH<7.4 also reduces the Hb-O_2 affinity. This makes physiologic sense since more oxygen will be released from Hb to the cells in the environment of lower pH. Thus, protons and CO_2 have the same triangle relationship with hemoglobin. One can view deoxygenated Hb as having more affinity for H^+ and $CO_{2,}$ which accounts for Haldane's effect(or oxyhemoglobin binds less to $CO_2/H^+)_.$ From the perspective of Mr. Bohr, both CO_2 and H^+ binding to hemoglobin reduces the affinity of Hb for oxygen and unloads more oxygen to the cells. The student could draw a creative cartoon of

Mr. Bohr and Mr. Haldane travelling in the red capsule and discussing or arguing about whose effects comes first!

The student should be careful to use oxygen language specifically. Changes in partial pressure of O_2 do not affect the Hb-O_2 affinity. It is the Hb-O_2 saturation which is determined by the PO_2 not the affinity of Hb for O_2. But, factors other than PO_2 (which is the variable along the x-axis of Hb-O_2 dissociation curve) affect the Hb-O_2 interactions.

For Bohr's effect, changes in PCO_2 and pH alter the interactions or affinity between Hb and oxygen.

8. What does the 50 in the P_{50} oxyhemoglobin index indicate?

Answer The 50 means 50 % Hb-O_2 saturation. The P_{50} is the partial pressure of O_2 that is *required* to give a 50 % Hb-O_2 saturation.

Concept In oxygen transport language, the P_{50} index is an indicator of the Hb-O_2 affinity. If the partial pressure *needed* to achieve 50 % Hb-O_2 saturation is higher, this mean that the hemoglobin has a reduced affinity for O_2. If you need more effort or pressure to make two persons like each other, then the natural chemistry between them must be low!

Normally, the P_{50} index has a value of 27 mmHg. At 27 mmHg, there will be a 50 % Hb-O_2 saturation. This P_{50} value is defined at certain set conditions. The conditions are body temperature 37 °C, blood pH 7.4, and blood PCO_2 40 mmHg. These are the conditions of arterial blood. Implied is that changes in body temperature, blood pH, or PCO_2 will change the P_{50} meaning, and they will change the affinity of hemoglobin for O_2.

If we look at the Hb-O_2 dissociation graph, if the P_{50} is increased >27 mmHg, there would be a shift of the curve to the right. A decreased P_{50} of less than 27 mmHg will shift the curve to the left.

We can now combine three different ways of talking about changes in Hb-O_2 affinity. An increased affinity will have a lower P_{50} associated with aleftward shift of the Hb-O_2 curve.

Conversely, a decreased affinity will increase the P_{50} and shift the Hb-O_2 curve to the right.

Oxygen transport and delivery must be appreciated from the cell perspective. Yes, the tissues are "cellfish" and are only satisfied when they have their oxygen needs met. A decreased Hb-O_2 affinity benefits the cell since more oxygen is unloaded to them. Think of the rightward shift of the HB-O_2 curve as releasing more O_2 to the cells. You give with your generous right hand!

9. What two factors affect the change in P_{50} during physical activity?

Answer An increased tissue PCO_2 and decreased pH increase the P_{50}, thus favoring the unloading of oxygen to the cells by decreasing the Hb-O_2 affinity.

Concept The body is well designed to serve its own diverse functions. During exercise, the blood flow to skeletal muscle increases as a result of increased cardiac output and local arteriolar vasodilatation in the muscles.

The oxygen diffusion at the microcirculation is also higher since the partial pressure gradient is greater. This PO_2 gradient is not due to any increase in arterial PO_2 (in moderate exercise, the PO_2 is unchanged). It is rather the decreased tissue $PO_2 < 40$ mmHg as a result of more cellular usage of oxygen. The surface area for diffusion is also increased by the vasodilation. Vasodilation also has the effect of decreasing the diffusion distance if some capillaries that are not so patent at rest are recruited during muscle activity.

In addition, the local conditions of more active cells all have the effects of promoting more unloading of O_2 from capillary blood. This is a most wonderful, appropriate physiologic arrangement in response to the need for more O_2. The three conditions—local PCO_2 (at 46 mmHg or higher), lower pH as well as warmer tissues during physical activity all loosen the affinity of Hb for O_2.

At rest, the tissue extraction of oxygen is about 25 %. This means that the total blood oxygen content is reduced by a quarter, from arterial 20 ml to 15 ml oxygen percentage. Since most of oxygen content in blood is hemoglobin bound, the percentage saturation of Hb-O_2 also decreases by about 25 %, from ~98 to 75 % (PO_2 at 40 mmHg). The normal consumption of oxygen is calculated by multiplying the cardiac output (5000 ml/min) with the tissue extraction of 5 ml O_2/100 ml of blood. This gives a value of 250 ml O_2/min.

During muscle activity, the venous blood PO_2 can be less than 40 mmHg. This means the Hb-O_2 saturation in venous blood will also be less than 75 %.

10. How does the change in P_{50} during carbon monoxide overdose worsen its toxicity?

Answer Carbon monoxide toxicity makes the hemoglobin bind to oxygen more tightly (reduced $P_{50)}$ and prevents oxygen unloading to the hypoxic cells.

Concept Hemoglobin has a 200 fold greater affinity for carbon monoxide (CO) compared to for oxygen. Thus, the blood oxygen content is drastically reduced in CO poisoning. Because CO has no smell, this makes it particularly dangerous as the presence of the gas is not noticed. The CO in blood does not affect the dissolved oxygen and the PO_2 is unchanged. The arterial chemoreceptors are, thus, not stimulated, and there is no increased breathing to indicate the infiltration of CO into the body. Perhaps, this is why CO inhalation is a common method of choice for persons who decided to end their lives in a relatively comfortable way.

In addition, the remaining oxygen in the Hb-CO saturated blood is less readily released to the tissues. This is due to an unusual change in the Hb-CO that increases the affinity of Hb for oxygen. This increased Hb-O_2 affinity is pointless in CO toxicity, when the most of the hemoglobin is occupied with CO instead of O_2. The situation is worsened since the remaining oxygen cannot be easily unloaded to the hypoxic cells in this severe category of anemic hypoxia. The CO-Hb–O_2 curve shows a shift to the left.

In a different situation that occurs at extreme high altitude, the respiratory alkalosis has a compensatory effect in increasing Hb-O_2 affinity. As the very low partial pressure of oxygen in the thin atmospheric air at the high altitude, this increased affinity helps the climber to improve the lung oxygenation. This is seen as a decreased P_{50} with a left shift of the Hb-O_2 curve.

At moderate altitudes, the rise in erythrocyte 2, 3 diphosphoglycerate (DPG) due to altitude hypoxia normally decreases the Hb-O_2 affinity (right shift of Hb-O_2 curve) to compensate and improve cell O_2 supply.

Chapter 9
CO_2 Respiratory Physiology

Carbon dioxide is generally viewed as a metabolic gaseous waste product. It is a major byproduct of cellular metabolism of energy substrate in our foods, including carbohydrates, fats, and proteins. The lungs are the site of elimination of CO_2, expired in our breath. Above this metabolic picture, the student should remember that CO_2 and its associated carbonic acid is a determinant of blood pH. In other words, pulmonary function is tightly linked to the homeostatic control of blood pH. In addition, the metabolite CO_2 is a vasodilator which regulates regional blood flow.

1. What is the major form of carbon dioxide transported in *arterial* blood?

Answer The partial pressure of CO_2 in arterial blood is 40 mmHg and in venous blood is 46 mmHg. In terms of blood content, the arterial CO_2 content is 48 ml CO_2/deciliter and 52 ml CO_2/dl in venous blood at rest. This means that for every 100 ml of pulmonary blood flow that passes the lungs, 4 ml of the 52 ml CO_2 is expired. With a resting cardiac output of 5 L/min, the rate of CO_2 released in the lungs is 200 ml/min of CO_2.

In both venous and arterial blood, CO_2 is transported in three major forms. A small percentage of CO_2 is dissolved in plasma and cytosol of red cells. A slightly larger portion is bound to hemoglobin to form carbamino-hemoglobin. The largest fraction (~70 %) of blood CO_2 content is in the form of plasma bicarbonate anion, HCO_3. Bicarbonate is formed from carbonic acid which is generated within the erythrocytes by hydration of CO_2. At the tissues, the CO_2 diffuses easily into the blood and into red cells down its partial pressure gradient. Thus, hemoglobin in red cells has three physiologic roles; besides the well-known fact of binding oxygen, hemoglobin also combines with CO_2 to form carbamino-Hb, and Hb is also an important blood buffer for pH control. Note that the three forms of CO_2 transported in blood are in equilibrium with each other; the dissolved, the Hb-bound and the plasma bicarbonate.

2. What enzyme function in red blood cells is essential for CO_2 transport?

Answer The erythrocyte has the enzyme carbonic anhydrase that converts CO_2 to carbonic acid, which then dissociates into the anion bicarbonate and hydrogen ion.

H. M. Cheng, *Physiology Question-Based Learning*, DOI 10.1007/978-3-319-12790-3_9

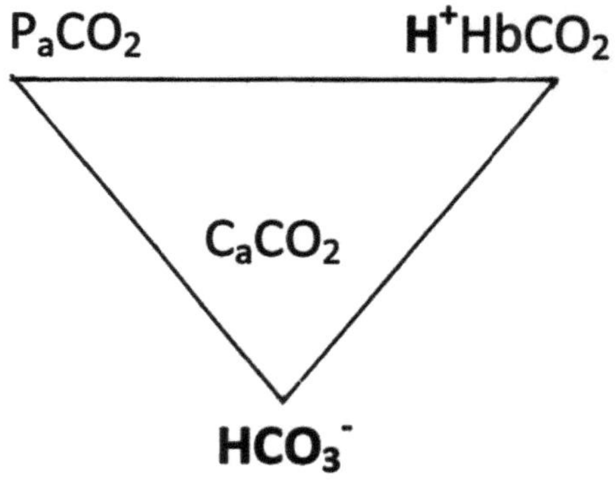

Arterial blood CO_2 content ($CaCO_2$) is distributed as three forms. Around two thirds are transported as converted CO_2, plasma bicarbonate. Smaller amounts are carried as carbamino-hemoglobin (Hb-CO_2) and as dissolved in blood. Hb also buffers the hydrogen ions generated from hydration of CO_2 by the red cell enzyme, carbonic anhydrase

Concept The red cells are essential in transporting CO_2 in blood, besides carrying oxygen. The red cells contain the enzyme carbonic anhydrase C@ which is needed for the hydration of CO_2 inside the red cells. The subsequent carbonic acid dissociation is the source of the major form of total CO_2 transported in the blood (bicarbonate anions). The reaction of carbon dioxide with water by C@ is a reversible reaction: $(CO_2 + H_2O <> H_2CO_3 <> H^+ + HCO_3)$. The reaction is driven to the right at the tissues. In the lungs, a reversal of the kinetics favors a leftward reaction with the release of CO_2 from the pulmonary capillary blood into the alveoli.

There are three other organs, in which the carbonic anhydrase enzyme plays a central role. The parietal cell of the stomach secretes hydrochloric acid. The proton that is secreted is generated inside the gastric parietal cells, catalyzed by a C@. In the pancreas, the ductal cells secrete bicarbonate which is a major component of pancreatic excocrine secretion. The ductal cells produce the bicarbonate by an action of C@ in the ductal cells. In the kidneys, both the reabsorption of filtered bicarbonate, tubular synthesis of bicarbonate, and tubular secretion of protons are effected by C@ present in the epithelial cells of specific nephron segments.

3. How is hemoglobin buffering action involved in CO_2 transport?

Answer Hemoglobin is an important blood buffer, an intracellular buffer circulating inside red blood cells! There are two reactions involving hemoglobin including the buffering action that enables the blood to carry excess CO_2 from the tissues to the lungs. The hemoglobin forms carbamino-Hb with CO_2 (do not confuse this Hb complex with a similar sounding carboxy-Hb which is formed during carbon monoxide toxicity).

Carbonic acid is generated inside red cells by the catalytic action of carbonic anhydrase. The dissociation of carbonic acid into bicarbonate and proton is enhanced by the buffering action of hemoglobin. By buffering H^+, the dissociation is promoted and more bicarbonate is produced. Bicarbonate, then, is exchanged at the red cell membrane for chloride and the increased biacarbonate concentration in venous blood represents the major form of CO_2 content in venous blood.

This bicarbonate/chloride interchange is reversed at the alveolar capillary, where the reaction is pushed towards release of gaseous CO_2 into the lungs. The red cell membrane chloride shift event has been called "Hamburger" effect, after the Dutch physiologist Hartog Jakob Hamburger.

4. State the effect of Haldane's effect at the muscle tissue during physical activity?

Answer Decreased partial pressure of oxygen at the exercising muscles will favour the uptake of carbon dioxide into the capillary blood.

Concept Haldane's effect describes how changes in the partial pressure of oxygen will affect the CO_2-carrying capacity of blood. Increased PO_2 will lower and decreased PO_2 will raise blood CO_2 content. This inverse relation between PO_2 and blood CO_2 content named Haldane's effect is obviously advantageous and physiological. At the lungs when blood becomes oxygenated and PO_2 increase from 40 to 100 mmHg, the blood "holds" less CO_2 and this favours CO_2 elimination at the lungs. At the tissues, when oxygen is extracted from the blood, the deoxygenated blood with lower PO_2 (from 100 to 40 mmHg) will have a greater capacity to take up metabolic CO_2 from the cells.

The mechanism for Haldane's effect can be biochemically explained. Deoxygenated Hb has a higher affinity for CO_2 and hydrogen ions. Thus, more carbamino-Hb is formed. The greater buffering of H^+ will also promote more CO_2 transport in the form of the major species, bicarbonate ions (see Q3 above).

The student must not have the mental picture that CO_2 will replace all the oxygen on hemoglobin. The venous deoxygenated Hb is still 75 % saturated with oxygen at rest. Thus, the carbamino-Hb and oxy-Hb coexist on the same carrier molecule. Similarly at the lungs, it should not be imagined that the oxygenation of blood will displace all the CO_2 from hemoglobin. In arterial blood, PCO_2 has a value of 40 mmHg, just 6 mmHg lower than in mixed, venous blood. The arterial blood CO_2 content has all three forms of CO_2—the bicarbonate, the carbamino-Hb, and dissolved CO_2.

You might say that CO_2 and O_2 have a love –"Hbate" relationship. The two respiratory gases are together on the hemoglobin molecule, and they also nudge each other away at specific places and at times of their life journey together!

5. What are the parameters of the *y*- and *x*-axes of the CO_2 transport curve?

Answer The *x*-axis is partial pressure of CO_2 and the *y*-axis is the blood oxygen content.

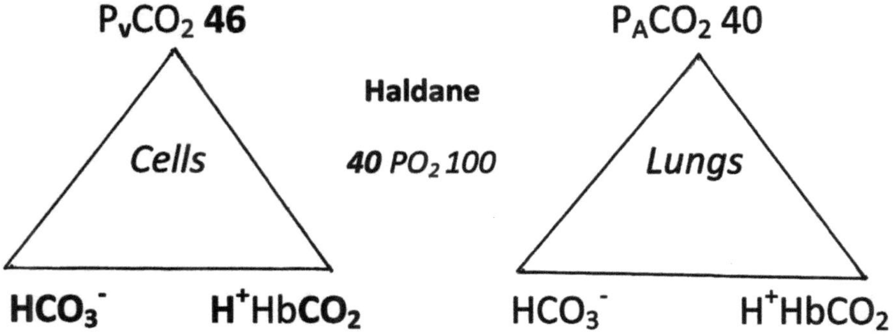

Carbon dioxide exchange at the cells and the lungs are dependent on the CO_2 partial pressure gradient at the metabolic and ventilatory sites, respectively. The concurrent change in the oxygen partial pressures at the cells or tissues also promote CO_2 extraction from tissues and CO_2 removal from pulmonary blood (Haldane's effect). The partial pressure of CO_2 is in chemical equilibrium with the hemoglobin-bound CO_2/H^+ and with plasma bicarbonate anions

Concept A graphical illustration of physiological relationships is helpful, and they not just show a single situation cause and effect (c & e) but a range of varying situations. Generally, a considerable number of students in a class will be averse to graphs. I call this "graph's disease/syndrome"!

By noting carefully the specific parameters of the axes, the student can become more versatile with understanding physiology in changing situations. In the CO_2

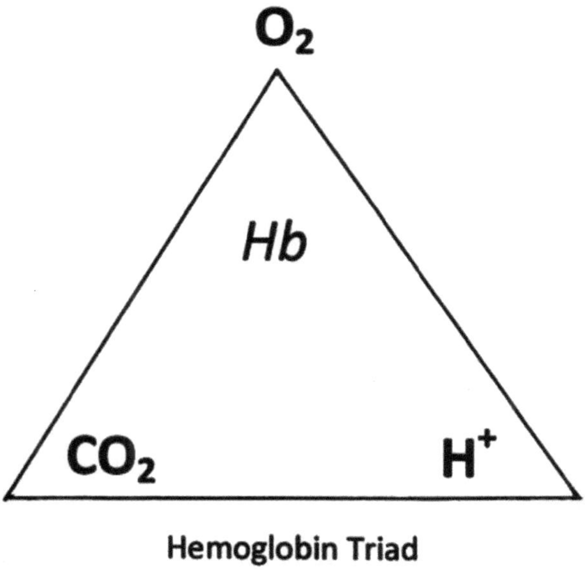

The red cell protein, hemoglobin has three carrier functions. Oxy-hemoglobin represents at least 99 % of blood oxygen content. Carbamino-Hb is a small component of blood CO_2 content. Hemoglobin is also an essential blood buffer, and this role of Hb prevents venous blood pH from becoming acidic. The buffering action of Hb for hydrogen protons also enhances the formation of bicarbonate anions as the major form of blood CO_2 content. Hb also carries nitric oxide (NO). DeoxyHb has less affinity for NO, so at the tissues NO is released to produce vasodilation transport curve, the *x*-axis is the partial pressure of CO_2 and the physiologic range of interest is normally shown in textbook as 30 to 50 mmHg. This range will include the 40–46 mmHg fluctuations for CO_2 in arterial-venous blood. Note that there is no subscript after the *P* for CO_2 in the axis. The *y*-axis is the blood CO_2 content expressed as milliliter CO_2/100 ml of blood. The *y* values shown in CO_2 transport graph are normally 40–60 as this will include the 48–52 ml CO_2/dl in arterial and venous blood, respectively.

There will always be two curves in the CO_2 transport graph. This is because the CO_2 content is affected by ongoing changes in the partial pressure of oxygen. If PO_2 is not considered, we only look at one of the curve. For example, the upper curve is given for when PO_2 is 40 mmHg (deoxygenated venous blood). This curve shows increasing CO_2 content with increasing PCO_2. A difference with the hemoglobin-oxygen transport curve is that there is no apparent plateau phase as in the case with Hb-O_2 when hemoglobin near saturation is reached.

In reality, at the lungs when PCO_2 decreases from 46 to 40 mmHg, the partial pressure for oxygen does not remain at 40 mmHg, but the lung oxygenation rapidly arterialized the blood to PO_2 ~100 mmHg. Thus, when following the events for CO_2 transport from tissues to the lungs, we have to change the curve from the upper one to the lower one where PO_2 has become 100 mmHg. Looking at the point of intersection at PCO_2 40 mmHg (*x*-value) for the two curves, it is clear that the CO_2 content (y values) is reduced more for the lower curve of oxygenated blood at PO_2 100 mmHg.

The presence of the two curves is a demonstration of the Haldane's effect. Haldane's effect is the effect of PO_2 on blood CO_2 content. This is described as a shift of the CO_2 transport curve to the right with higher PO_2, i.e., at any given PCO_2, oxygenation (at the lungs) will further decrease the CO_2-carrying capacity of blood and reduce its CO_2 content.

6. How is the pH of venous blood related to the concentration of bicarbonate?

Answer The pH of venous blood is dependent on the ratio of bicarbonate to the carbonic acid concentration.

Concept There are several chemical buffers in the extracellular fluid (ECF) which includes the blood volume. These chemical buffers provide the first line of defence against a pH threat. The main chemical buffer in the ECF is the bicarbonate/carbonic acid buffer. At any pH, the base and acid components of all the different chemical buffers are linked to each of the other chemical buffers by the same pH (isohydric principle). We need only, therefore, to monitor the bicarbonate and carbonic acid concentrations to ascertain the pH landscape. The carbonic acid concentration is determined by the partial pressure of CO_2 and at the normal arterial blood PCO_2 of 40 mmHg, and pH of 7.4, the carbonic acid concentration is 1.2 mmol (a conversion factor of 0.3).

Normal bicarbonate concentration is arterial blood is 24 mmol/L. The ratio is then 20 at the pH of 7.4. The bicarbonate/PCO_2 buffer system is the main buffer as the buffer pair is an "open" system. The "open" term indicate that the amount/concentration of the HCO_3 and PCO_2 is not limited. Both the base and acid components are "open" to be increased or decreased in compensatory response to pH changes. The bicarbonate is "open" to be modified by the renal function and the PCO_2 is "open" to the respiratory function. An elegant relationship for this "open" concept of pH control by two major organs acting via the HCO_3/PCO_2 is

$$\text{pH} \sim [\text{HCO}_3]/\text{PCO}_2 \quad \ldots\ldots\ldots \quad \text{pH} \sim \text{kidneys/lungs}$$

The venous pH is slightly acidic due to the effective buffering of protons by hemoglobin. The venous bicarbonate base concentration is higher than that in arterial blood since a large fraction of metabolic CO_2 has been transformed to the anion. However, the venous pH is not due to the absolute concentration of bicarbonate but to the ratio of HCO_3 to PCO_2. The PCO_2 in venous blood is 46 mmHg, up from 40 mmHg in arterial blood.

7. How are the three parameters of alveolar ventilation equation related?

Answer The partial pressure of CO_2 in alveolar air (P_ACO_2) is related to the ratio of the rate of CO_2 production (VCO_2) to the alveolar ventilation (V_A).

Concept At rest, if the rate of CO_2 production is unchanged, the PCO_2 is inversely related to the rate of alveolar ventilation. If this is plotted on a graph, there is a hyperbolic inverse relationship between alveolar ventilation (x-axis) and alveolar PCO_2 (y-axis). The value of resting alveolar ventilation should then fall on the graph at PCO_2 of 40 mmHg. If a person voluntarily hyperventilates, the alveolar PCO_2 will decrease. Since arterial blood PCO_2 is the same as alveolar air PCO_2, the blood will be hypocapnic, and the condition is also called respiratory alkalosis since hypocapnia will increase the blood pH.

What will happen if the rate of CO_2 production is increased? How will the relationship between alveolar ventilation and alveolar PCO_2 change? The basic hyperbolic, inverse relationship between the ventilation/PCO_2 will remain the same. However, for the alveolar air PCO_2 to remain at 40 mmHg (y-axis) at the greater rate of metabolic CO_2 release from cells, the alveolar ventilation has to increase (y-axis). Thus, the hyperbolic curve shifts to the right with a higher cellular CO_2 generation.

This is a useful place to consider cause and effect (c and e) mechanism in physiology. Voluntary hyperventilation as a cause will lead to decreased PCO_2. If increased PCO_2 is the cause, then it will stimulate increased ventilation. In comparing these two situations, the hyperventilation is the cause in the former and the effect or result in the latter. Similarly, any case of hypoventilation will cause an elevation in the alveolar/arterial PCO_2 (hypercapnia). Hypoventilation can also be an effect of hypocapnia. Distinguishing the cause/effect (chicken nor egg?!) in every scenario will help the student to understand homeostasis, compensatory feedback in Physiology.

8. How does CO_2 participate in two ways to ensure adequate oxygen supply to exercising muscles?

Answer The local increase in tissue CO_2 helps to unload more oxygen from blood. The local CO_2 also vasodilates the blood vessels to increase the perfusion to the exercising muscles.

Concept Carbon dioxide is the major metabolic byproduct from the cellular processing of energy substrates namely carbohydrates, lipids, proteins. Increased active muscle contraction needs more oxygen and produce more CO_2.

The increased supply of oxygen is partly contributed by increased alveolar ventilation. In moderate exercise, however, relatively constant arterial PCO_2 cannot account for the stimulation of a sustained hyperventilation during exercise.

However, at the tissue level, CO_2 has definite actions that increase both the unloading of oxygen to the cells and the increased blood tissue perfusion. The local tissue PCO_2 is higher than 40 mmHg. Blood that are exposed to this higher PCO_2 will have a decreased hemoglobin affinity for oxygen. Hemoglobin "loosens up" and release more O_2 to the cells. This is called Bohr's effect which is also observed when the local tissue pH is less than 7.4.

Metabolite vasodilators also include CO_2. The arteriolar smooth muscle relaxes and vasodilates in response to local tissue hypercapnia. The skeletal muscle blood flow is increased in proportion to the muscle cellular metabolism.

Thus, we have an example of a "physio-synthesis" involving CO_2! The student can train herself to have a mental habit of integrating, associating, and finding linkages between different situations that involve the same molecule or same physiologic principle or event like in membrane transportation.

It is also a fascinating "coincidence" that besides CO_2, all other metabolites that have a vascular effect are vasodilators in the respective tissue, e.g., the adenosine in coronary circulation. The student can look out and discover for more organizing patterns and mechanistic design in physiology.

9. How will an enlarged functional residual capacity (FRC) affect the alveolar air CO_2?

Answer The alveolar air partial pressure of CO_2 will increase above 40 mmHg.

Concept FRC is the air in the lungs at the end of a normal resting expiration. The inspired tidal volume of the next breath will mix with this FRC en route to the alveolar gas exchange region of the lungs. Carbon dioxide diffuses into the alveoli

My Bohr's is good for Oxy-gen

My Bohr's Unload O2 to cells

CO_2, Oxygen Reduction

Low pH also Bohr's can you see?

Feed back O Feed back

Bohr's reduce Hb affi-nity

Feed back O Feedback

High P50 is good for me!

Physio-lyrics to popular tunes. This composition highlights the beneficial effects of tissue CO_2 and lower pH in unloading of oxygen to the cells. The CO_2/pH Bohr's effect reduces the hemoglobin-O_2 affinity, and from the cell's perspective, this has a positive effect on the cellular O_2 supply

since there is a PCO_2 partial pressure gradient between pulmonary blood (entering the alveoli at 46 mmHg) and the alveolar air (controlled at 40 mmHg).

One can picture that the alveolar air PCO_2 is an equilibrium value, a balance between the release of CO_2 from blood and the removal of CO_2 in the expired tidal volume. The FRC can be viewed as a buffer air region through which the expired CO_2-rich alveolar air and the inspired CO_2-poor air moves.

If the FRC is enlarged, a condition that develops in chronic obstructive lung disease, the abnormal increased air reservoir buffer will alter the alveolar air PCO_2. The balance of the air inflow and outflow will be changed to a higher alveolar air PCO_2 (simplistically, the student can see this as a larger FRC region for CO_2 to get out from). When this happens, the partial pressure gradient for optimal diffusion of CO_2 from pulmonary blood into alveolar air will be reduced.

FRC, the end-expiratory lung volume at rest, includes both the anatomical dead space and the alveolar respiratory space. The residual volume at the end of a forced expiration still includes the anatomical dead space and a reduced alveolar respiratory space/zone.

10. Why is the diffusion coefficient for CO_2 at the alveolar capillary membrane higher than for oxygen?

Answer The much higher solubility of CO_2 accounts for greater net diffusion coefficient for CO_2 through blood/air medium.

Concept CO_2 is a bigger molecule than oxygen. In air, the diffusion of oxygen is expected to be greater than for CO_2. However, when diffusion is considered between alveolar air medium and pulmonary blood, the solubilities for CO_2 and oxygen has to be factored in. Carbon dioxide is more soluble in blood. The amount of a gas dissolved in solution is described by Henry's law, which states that the dissolved gas content is dependent on the gas partial pressure and the solubility coefficient. For oxygen, it is 0.003 ml O_2/dL blood/mmHg, whereas it is much higher for CO_2 at 0.07 ml CO_2/dL blood/mmHg. Dissolve CO_2 is thus a higher fraction (5 %) of total blood CO_2 content than dissolved O_2 (2 %).

The diffusion coefficient for CO_2 is about 20 times higher than the diffusion coefficient for oxygen. For the same partial pressure gradient across the alveolar-capillary membrane, CO_2 diffuses 20 times faster than O_2. We can see this diffusive difference in the smaller partial pressure gradient for CO_2 (46/40 mmHg) than that needed for O_2 (100/40 mmHg).

The diffusion capacity for CO_2 across the alveolar-capillary membrane is governed by Fick's law of diffusion. In addition to the gas diffusion coefficient, the area available for diffusion and the thickness of the alveolar-capillary (a-c) membrane are also influencing factors.

In several occupational-related pulmonary disease, there is thickening of the a-c membrane. This will interfere with lung oxygenation and equilibration of oxygen may not be achieved in the short pulmonary capillary blood transit time. This "alveolar-capillary block" of gas diffusion is not as great a problem with CO_2 diffusion because of its greater diffusion coefficient. Thus, hypoxia and not hypercapnia is the key pathophysiology in "a-c" diffusion block.

Chapter 10
Respiratory Control

We breathe on an average 10–12 times/min. The volume of air we breathe in and out is also relatively constant, although you are unaware of the air movement when reading this page. When we are physically more active, both the frequency and the depth of each breath increase. How does our body maintain adequate ventilation at rest and step up the respiratory function during exercise?

1. Which is the major chemical that controls normal breathing?

Answer The main chemical regulator of respiration is carbon dioxide.

Concept Breathing to stay alive is naturally linked with breathing in oxygen from the air. The respiratory control mechanisms in the body in reality are more sensitive to the CO_2 in the extracellular fluid (ECF). Regulation of respiration is integrated by neurons in the brain stem which receives inputs from the peripheral arterial chemoreceptors. In the brain stem, there are also central chemoreceptors that sense CO_2 changes in the blood. There are a number of observations that account for the need for greater sensitivity to CO_2. The neurons are particularly affected by changes in pH of the ECF. An increased CO_2 produces an acidotic environment which depresses neuronal activity. For oxygen, even when the partial pressure is reduced to 60 mmHg from 100 mmHg, the hemoglobin-O_2 saturation is still about 90 %. Hypoxia only stimulates the peripheral chemoreceptors (carotid and aortic bodies), whereas for CO_2, both the central and peripheral receptors respond. Part of the chemoreceptor response to CO_2 is indirect via the generation of protons from carbonic acid. The central chemoreceptors within the brain stem are sheltered from hydrogen ion formed outside the blood brain barrier. However, CO_2 can diffuse into the brain interstitial space and be converted to H^+. The proton generated with the brain stem then stimulate the central chemoreceptors. Increased afferent inputs from both peripheral and central chemosensors to the respiratory control neurons in the brain stem will then increase both the tidal volume and the frequency of breathing.

2. What central nervous-system neurons produce increased depth but decrease frequency breathing in a person during voluntary action?

H. M. Cheng, *Physiology Question-Based Learning*, DOI 10.1007/978-3-319-12790-3_10

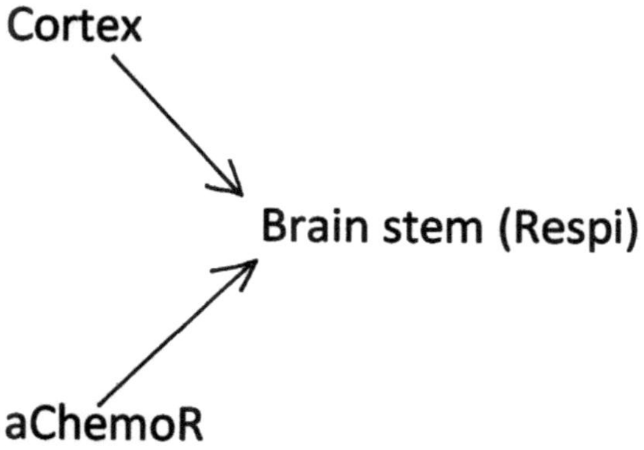

The integrated circuit of respiratory neurons in the brain stem generates the autorhythmic "pacemaker" activity that governs a respiratory cycle. These respi-neurons can be overrided or bypassed by cortical signals (e.g., voluntary hyperventilation of breath holding). The arterial peripheral chemoreceptors provide feedback sensory signals on pH, PO_2, and PCO_2 in blood. In the brain stem, there is another group of central chemoreceptors, in proximity to the respiratory neurons

Answer Voluntary control of the lung mechanics is via cortical neurons, and the motor efferents bypass the involuntary respiratory control neurons in the brain stem.

Concept Breathing is unique in that there are both an automatic as well as a voluntary control. You are not conscious of your breathing while reading this page; respiratory pacemaker neurons sets the resting depth and rate of breathing. You can hold your breath at will (you must go under water in a swimming pool!), or you can take a deep breath. When you do some deep, slow breathing exercises, the inspiratory skeletal muscles (diaphragm, external intercostals muscle) receive action potentials down their respective alpha motor neurons.

Clinically, a patient can sometime lose the functions of the respiratory brain stem neurons, either through physical trauma or disease. In such conditions, the person can still stay alive by a conscious effort in breathing. This is obviously exhausting. And, when the patient retires to sleep, she is put on a respirator. This restricted breathing, confined and sustained only by voluntary action, is called "Ondine's curse" after a tale, where the jealous Ondine punished her lover by taking away his normal breath since he became unfaithful and broke his vow to love Ondine "with his every breath"! Sighing and yawning are basically also involuntary reflexes, but they can also be voluntarily "imitated"!

3. How does voluntary hyperventilation increase the time a person can remain under water?

Answer Hyperventilation produce a hypocapnia that delays the time for the carbon dioxide to build up in the body to stimulate the respiratory center.

Concept This is another example of the greater sensitivity of the respiratory response to CO_2 compared to oxygen. The student may associate the voluntary

The peripheral arterial vascular chemo-sensors that serve respiratory control are sensitive to three blood parameters. Decreased partial pressure of O_2 (PO_2), increased PCO_2, and acidic blood pH will activate the carotid/aortic chemoreceptors to send stimulatory impulses to the respiratory neuronal center in the brain stem

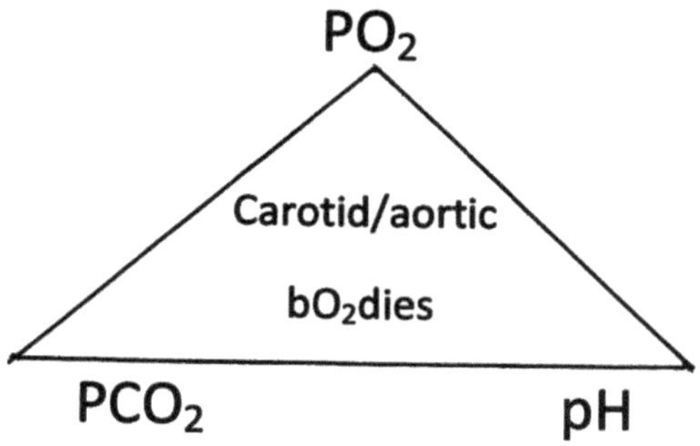

hyperventilation with more oxygenation of the blood and thus a longer submerged time under water. The "break-point" at which the person remains under water can no longer suppress the urge to breath which is actually due to the strong stimulus of increasing CO_2 in the ECF.

Quantitatively, the hyperventilation actually does not increase much the blood oxygen content. This is, because even at normal resting rate of breathing, the hemoglobin-oxygen saturation is 97 %, and Hb-bound O_2 makes up most of the oxygen content with a small percentage as dissolved oxygen. Does then the hyperventilation that accompanies exercise serve any physiologic role if increased oxygenation is not a primary benefit? In line with the question above, hyperventilation has an essential role in removing more CO_2 that is produced by more cellular metabolism during physical activity. The alveolar partial pressure of CO_2, PCO_2, is determined by the ratio of cellular CO_2 production and the alveolar ventilation as expressed in the alveolar ventilation equation: $P_ACO_2 = V_{CO_2}/V_A$. During moderate physical activity, the PCO_2 in alveolar air and thus the arterial blood is relatively unchanged. The higher rate of CO_2 production is balanced by the increase alveolar ventilation. The interesting question in exercise is then, "How is the hyperventilation stimulated if not by any increase in CO_2?" (see question 5 below).

4. Why does the prolonged, increased voluntary breathing produce dizziness in a person?

Answer Cerebral vasoconstriction due to the induced hypocapnia.

Concept Increased ventilation produces a respiratory alkalosis, since the CO_2 is "blown" off by the lungs. The cerebral arterioles are particularly sensitive to changes in blood CO_2. A reduced partial pressure of CO_2 will constrict the cerebral vessels. This lessens the cerebral blood flow and gives the sensation of giddiness.

This arteriolar response to CO_2 is also part of the intrinsic autoregulatory mechanism of cerebral blood flow. The cerebral vasculature has the inherent ability to maintain a relatively constant perfusion in spite of blood pressure fluctuations over a certain autoregulatory range (60–160 mmHg). If there is an acute decrease in cerebral flow, soon the local CO_2 in the brain tissues builds up. The cerebral arterioles will then vasodilate and blood flow is sustained. This is an example of a physiologic role of CO_2, a gas which should be viewed more than just a metabolic end product that needs to be eliminated by the lungs during expiration.

If the decrease in cerebral blood flow is due to hypovolemia/hypotension, it is useful to contrast the local compensatory cerebral vasodilation with the need to sustain an adequate head-driving arterial blood pressure. Concurrent with the reduced cerebral vascular resistance, there will be a systemic increase in the total peripheral resistance (TPR). The compensatory increase in TPR is effected by baroreflex activated sympathetic nerve action on selective arteriolar vasoconstriction (including skin, splanchnic, renal).

5. How do the blood pH and the CO_2 change during heavy exercise?

Answer There will be a decrease in blood CO_2 due to stimulation of respiration by lactic acidosis.

Concept During voluntary hyperventilation (question 3 above), there can be an increased in arterial blood PO_2 and a respiratory alkalosis due to a decreased in PCO_2. Surprisingly, in moderate exercise like jogging, the arterial blood PO_2 and PCO_2 remain unchanged. How then is exercise hyperventilation stimulated and sustained? There are evidences that muscle afferent mechano- as well as muscle chemoreceptors provide some of the stimulatory inputs into the respiratory neurons in the brain stem. There could also be, not easily detectable by conventional measurements, rapid fluctuations in the blood gases that maintain the hyperventilatory stimulus.

The situation during heavy exercise is different from that in moderate physical activity. The imbalance between cellular metabolic needs and the oxygen supply soon result in a lactic acidosis. This metabolic acidosis will now stimulate the peripheral chemoreceptors to increase alveolar ventilation. The arterial blood PCO_2 will become less than 40 mmHg. The compensated blood pH by the lower PCO_2 will still be acidic as lactic acidosis is the primary cause. Blood pH is determined and reflected by the ratio between the chemical buffer, bicarbonate/carbonic acid base/acid component pair. Lacticacidosis will reduce the bicarbonate and the compensatory ventilation will decrease the carbonic acid that is formed by the hydration of CO_2.

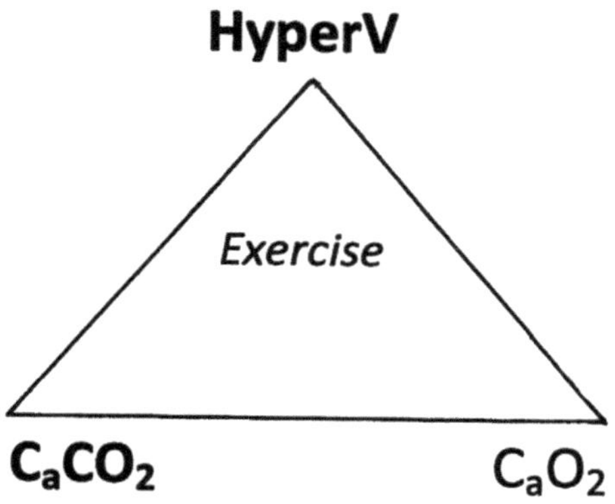

The increased alveolar ventilation during physical activity is primarily for eliminating CO_2 from the ECF/blood rather than to increased blood oxygen content. Accumulation of CO_2 will make the ECF/blood acidic, and neuronal functions are suppressed by acidosis. Increased rate of O_2 delivery to cells are effected by a greater cardiac output rather than by lung oxygenation since a higher ventilation rate does not markedly increase oxygenation to above normal blood O_2 content

6. How do blood pH and CO_2 change during ascent to high altitude?

Answer Hypoxia induced hyperventilation produces a respiratory alkalosis.

Concept Exposure to low-inspired oxygen concentration in the atmospheric air will lead to stimulation of alveolar ventilation. Do not imagine like some students that the greater physical effort of climbing is the predominant stimulus for the increased breathing (keep in mind that the ascent is a slow, enjoyable one generally!). The compensatory hyperventilation produces hypocapnia which alkalines the blood. The respiratory alkalosis opposes the hypoxic stimulation since CO_2 is the more potent chemical regulator of respiration. Thus, it will take several days for the person to acclimatize to high altitude. Part of the acclimatization process involves excretion to increase bicarbonate in the urine. This compensatory renal event will help to decrease the pH of the brain interstitial fluid that bathes the central chemoreceptors. Bicarbonate is the main base in the ECF and a component of the major chemical buffer system, bicarbonate/carbonic acid in ECF.

Mountain enthusiasts are prescribed a drug that promotes urinary bicarbonate excretion to hasten acclimatization. This drug is an inhibitor of the enzyme carbonic anhydrase C@. The C@ action at the nephron is essential for the normal reabsorption of filtered bicarbonate.

Note in this mountain top scenario, the blood PO_2 is lower than 100 mmHg (decreasing value with increasing altitude). The blood PCO_2 is also lower than 40 mmHg due to the compensatory hyperventilation.

Compare the situation in heavy exercise at sea level—the blood PO_2 can be slightly increased, and this is accompanied by a reduced blood PCO_2 due to the lactic acidosis stimulation of breathing.

Local changes in alveolar air PO_2 or PCO_2 will act to optimize ventilation/perfusion matching. In underventilated alveoli, the decreased alveolar PO_2 will produce hypoxic pulmonary vasoconstriction (HPV). If the alveolar PO_2 is all low, as at high altitude, this HPV can lead to pulmonary hypertension. In underperfused alveoli, the lower alveolar PCO_2 will bronchoconstrict to shunt airflow to better perfused alveoli. Note and distinguish these PO_2 and PCO_2 lung effects from the arterial and central chemoreceptor effects to blood PO_2, PCO_2 changes

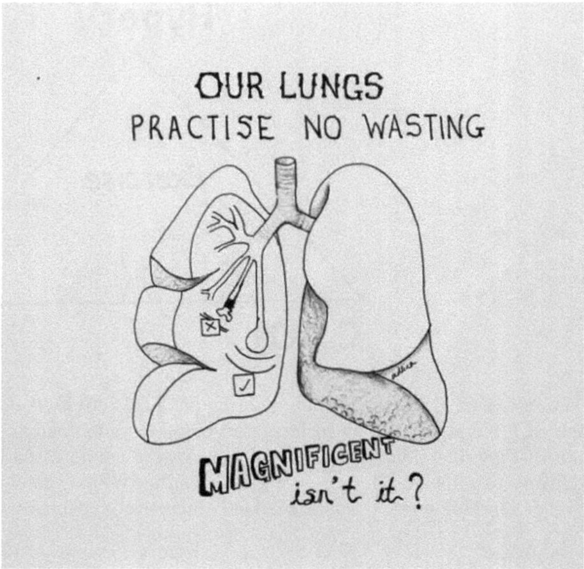

7. How does decreased oxygen supply to the brain produce a reflex protective cardiovascular response?

Answer The central ischemic response is due to the direct activation of the cardiovascular control neurons in the brain stem.

Concept Hypoxia does not stimulate the central chemoreceptors but only the peripheral arterial chemoreceptors at the aortic and carotid bodies. Generally, the respiratory neurons in the brain stem receive major inputs from the chemoreceptors. The cardiovascular regulatory neurons are also located in the brain stem. In situations when there is threatened ischemia of the brain, hypoxia can activate a direct emergency reflex response from the brainstem cardiac and vasomotor neurons. This is logically linked to, e.g., hypovolemia/hypotension as a result of blood volume loss. In order to maintain cerebral perfusion, the hypoxia acting on the brain stem produces a compensatory reflex similar to the baroreflex that responds to the hypotension. Tachycardia and increased TPR are part of this central ischemic response that aimed to elevate the arterial blood pressure to life-sustaining value.

In the case of a reduced cerebral blood flow that is caused by a "head" factor, i.e., increased intracranial pressure, a similar central ischemic reflex will be triggered. Interestingly, the increased blood pressure (above normal) will then result in a baroreflex bradycardia in this case of a mechanical compression of cerebral vessels by fluid pressure in the closed brain box. This hypertension/bradycardia paired events resulting from compensatory feedback mechanisms to the elevated intracranial pressure is also called "Cushing reflex."

8. Compare the venous blood PO_2 in stagnant and histotoxic hypoxia.

Answer There will be a decreased PO_2 instagnant hypoxia and an increased PO_2 in hitotoxic hypoxia.

Concept Hypoxia should be viewed from the perspective of the cell. The cell is the consumer target of the function of the cardiorespiratory system in terms of oxygen delivery. As long as the cells are not able to have enough oxygen for its metabolic needs, a hypoxic condition is present. In stagnant hypoxia, the oxygenation is normal at the lungs. However, an inadequate blood flow cannot supply the cells although the blood oxygen content is normal. This, e.g., occurs if the heart fails and cannot pump sufficiently to maintain a normal cardiac output. The rate of oxygen delivery milliliter O_2/min is equal to the oxygen content x of the cardiac output. A longer transit time during stagnant hypoxia at the tissues will permit the cells to extract more oxygen from the capillary blood. However, the unloading of O_2 to the cells is still limited by the partial pressure gradient. A fresh inflow of oxygenated arterial blood with PO_2 of ~100 mmHg is essential to ensure a PO_2 gradient for oxygen diffusion and uptake by the cells. The venous PO_2 will be less than the usual 40 mmHg.

The arterial blood PO_2 is unchanged in stagnant hypoxia. This is also the case in histotoxic hypoxia. The hypoxia is not due to oxygenation or any problem with peripheral blood perfusion. The cells are prevented from using the oxygen due to, e.g., metabolic inhibitors. Although there is a partial pressure gradient between capillary blood and the interstitial fluid, no net oxygen diffuses since the cells are not using the O_2. The blood picture characteristic of histotoxic hypoxia is an elevation of venous blood PO_2 markedly above 40 mmHg. If the cells are completely deprived of O_2, the venous PO_2 would theoretically be unchanged from its arterial PO_2 value (of course, the patient would be dead and you would not be asking the laboratory to measure the venous PO_2!).

9. Why is high oxygen therapy for chronic pulmonary conditions not recommended?

Answer Sensory adaptation of the chemoreceptors to chronic hypercapnia results in the need for a hypoxic drive to maintain breathing and keep alive.

Concept Sensory adaptation occurs in receptor mechanisms although some receptors adapt rapidly and some slowly. Pain or nociceptors adapt slowly and this obviously has a protective role to inform us of the presence of any lingering tissue injury. For touch/pressure cutaneous receptors, they adapt, and we sometimes find ourselves going into the shower with our spectacles on, our facial receptors adapted to the visual device.

In a patient with long-standing pulmonary disease, the blood PCO_2 is slightly elevated most of the time due to the subnormal alveolar ventilation. The chemoreceptors, exposed and stimulated by the chronic hypercapnia will eventually adapt and become no longer responsive to stimulate and maintain ventilation. What develops in the patient is a switch to a hypoxic drive. A lower than normal PO_2 in the chronic pulmonary dysfunction is now the only driving stimulus that ensures an adequate alveolar ventilation. This naturally leads us to see why a high oxygen inhalation may not be helpful but could be detrimental to the patient's breathing. Restoring the blood PO_2 also removes the only available hypoxic drive of ventilation. The reduced ventilation will then be insufficient to eliminate CO_2, which will accumulate in the blood. A progressively severe respiratory acidosis can result.

10. How does the chemoreceptor produce a release of "neurotransmitter" onto the afferent nerve that inputs to the respiratory center?

Answer The peripheral chemoreceptor cells senses hypoxia, in particular, the decrease of partial pressure of oxygen in the arterial blood that perfuses pass the carotid and aortic bodies. There is an ionic mechanism that translates the hypoxic sensing to an action potential in the afferent nerve fiber that ends on the chemoreceptor cells. In chemoreceptor, cells that are potassium channels are sensitive to oxygen. Lack of oxygen in the chemoreceptors decreases the ionic conductance of these potassium channels. The reduced K^+ efflux leads to depolarization of the chemoreceptor cells. Calcium influx via voltage-gated channels and a chemical transmitter is released and stimulate the afferent fiber that supplies the chemoreceptors.

This hypoxia induced depolarization effect is also seen in the mechanism of hypoxic pulmonary vasoconstriction (HPV). The pulmonary smooth muscles are depolarized by hypoxia and contract. This unique HPV has a role in optimizing ventilation/perfusion matching throughout the lung.

This contrasts with the systemic arteries, which have potassium channels that are closed by adenosine tri-phosphate (ATP). Hypoxia will reduce cytosolic ATP, and this opens the K^+ channels. Increased potassium cation efflux hyperpolarizes the systemic arteriolar smooth muscles and produces vasorelaxation. Arteriolar vasodilation is part of the compensatory mechanism at regional blood flow in the systemic circulation including in autoregulation, active, and reactive hyperemia.

Part III
Renal Physiology

Introduction: Renal Physiology

The kidneys produce urine. Pee Wee! The kidneys are not merely excreting urine and its contents as a waste product. The formation of urine is linked to a diverse range of physiological functions. The ability of the kidneys to vary the urine concentration is part and parcel of renal osmoregulation. The excretion of small or large urine volumes is associated with regulation of water balance in the body. Renal osmoregulation and control of water balance are both linked to the homeostasis of the common parameter of extracellular fluid (ECF) sodium concentration.

The ECF volume and blood volume are also under the governance of the kidneys. ECF volume is determined by the total body sodium, the cation being the key extracellular osmoactive electrolyte. The plasma volume, a fourth part of ECF is thus under renal control. The kidneys also secrete an erythropoietic hormone that maintains the normal hematocrit. Maintaining normal total body sodium or sodium balance is a major role of the renal nephrons. The nephrons and its supply of blood vessels are the target of renal sympathetic nerve which acts during sodium conservation.

It might seem odd, unrelated, and a surprise to think about a sympathetic neural activity being involved in sodium electrolyte balance. The kidneys are the primary source of the hormone/enzyme renin which is the initiator of a family of antinatriuretic hormones including angiotensin II and aldosterone. Blood volume control by the kidneys is part of what is also termed "long-term" blood pressure (BP) regulation. To remind students not to forget this, think of BP and BPee.

The other important ECF cation that is under renal control is potassium. The renal handling of potassium includes tubular reabsorption and secretion. The adrenal steroid hormone aldosterone has a dual action in regulating the potassium balance besides sodium balance.

The blood pH cannot remain at the normal 7.4 if our kidneys malfunction. The renal tubules secrete protons, reabsorb and synthesize bicarbonate which is quantitatively the major base in the ECF. Think of this essential renal function in acid–base balance as peeH. There is a transmembrane exchange phenomenon between potassium and hydrogen at all cells, including at the renal tubular cells.

The kidneys never walk or wee alone! Renal functions are integrated with cardiovascular physiology in ensuring normal arterial blood pressure. The kidneys W (wee) and the lungs V (ventilation), closely function together in arterial blood pH control. The kidneys and lungs sequentially generate the vasoconstrictor circulating peptide, angiotensin II which besides increasing total peripheral resistance is also a central mediator of euvolemia.

Chapter 11
Renal Hemodynamics and GFR

The resting kidneys receive around 20% of the normal cardiac output. About 90% of this renal blood flow (RBF) enters the nephrons (estimated 1 million/kidney) to be filtered at the glomerulus. Filtration is a voluminous event, 125 ml/min or 180 L daily. In a 70 kg male adult, the total body fluid is 42 L of which extracellular fluid (ECF) volume is a third at 14 L. Thus, glomerular filtration processed almost 13 times the total ECF and reflects a major role of the kidneys in the homeostasis of the ECF, the "internal aqueous environment" that bathes all cells. Filtration is the first step in urine formation. Urine flow rate is about 2 L/day, highlighting a large amount of water reabsorption from the glomerular filtrate. The final urine that is excreted is the net output from the three basic renal handling processes for both water and solutes—filtration, reabsorption, and secretion.

Changes in RBF produce parallel changes in glomerular filtration rate (GFR). Thus, renal autoregulation of GFR, an essential first step in urine formation, is linked to autoregulation of RBF. The autoregulatory mechanisms of RBF (myogenic and macula densa sensing) are explained hemodynamically by the same "Flow = Pressure/Resistance" equation, the resistance altered being the preglomerular afferent arteriole. The glomerulus is a unique capillary in being sandwiched between two arterioles—the afferent and the efferent. Downstream from the glomerular capillary network separated by the efferent arteriole is the peritubular capillary, which participates in tubular reabsorption and secretion.

1. What two determinants are used to calculate the GFR?

Answer The GFR is determined by the product of the filtration coefficient (K_f) and the net glomerular filtration pressure (*nFP*).

Concept The filtration coefficient is dependent on two factors, the surface area available for filtration of the plasma water and the water or hydraulic permeability. The student should note that the GFR is not dependent on solute permeability as we are dealing with the movement of fluid volume not filtered solute load. The surface area for filtration can be altered by the degree of contraction of the glomerular mesangial cells. Vasoactive agents can reduce the K_f when the mesangial cells contract.

H. M. Cheng, *Physiology Question-Based Learning*, DOI 10.1007/978-3-319-12790-3_11

The glomerular filtration rate (*GFR*) is determined by the net filtration pressure (*nFP*). nFP is the balance of the three Starling's forces across the glomerular capillary; opposing hydrostatic pressures in the glomerulus and Bowman's capsule and the glomerular oncotic pressure

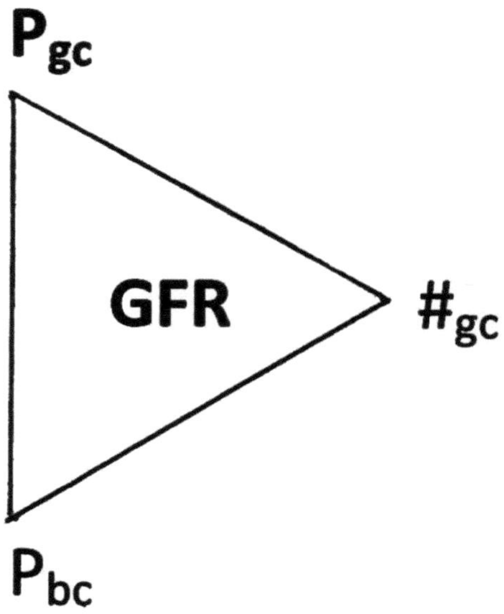

The glomerulus is a capillary, and so the Starling's capillary forces are operative in explaining glomerular filtration. If the students appreciate capillary dynamics, when this topic was covered under cardiovascular physiology, the derivation of the net filtration pressure is a wee (meaning as easy as our urine flow!)

In the glomerular capillary, the two Starling's forces are the hydrostatic pressure and the plasma oncotic pressure. In the Bowman's capsule (equivalent of the interstital space in other tissues), only the hydrostatic pressure is considered as little protein leaks out from the glomerulus. Hence, the oncotic pressure in the Bowman's space is near to zero mmHg.

There are then three Starlings—forces with the glomerular hydrostatic pressure being the only force promoting filtration. The net filtration pressure is the arithmetic sum of the three forces.

A few unique characteristic of the glomerular Starling's forces deserve mention. First, the glomerular hydrostatic pressure (Pgc) is distinctly higher than in other microcirculations. Second, the Pgc only drops slightly along the glomerulus. This relatively constant high hydrostatic pressure is obviously essential to produce a large GFR of 125 ml/min(or 180 L per day!). The presence of a postglomerular high resistance efferent arteriole sustains the Pgc for filtration.

The glomerular oncotic pressure (#gc), on the other hand, increases along the capillary. From about 25 mmHg at the afferent arteriolar end of the glomerulus, the #gc reaches about 40 mmHg "downstream." This increasing #gc is due to the high filtration fraction in the glomerulus, a value of 20 %, so the plasma protein is progressively more concentrated along the glomerulus. The value of the #gc in the GFR equation is therefore not a single value but a mean value.

There is no capillary reabsorption at the glomerulus, and the net filtration pressure decreases progressively along the capillary.

Normal resting RBF is around 20 % of cardiac output. Note that the filtration fraction is a portion of the renal plasma flow, not RBF as red cells are confined within the glomerulus.

2. What is the expected effect of sympathetic action of the afferent arteriole on filtration fraction?

Answer The filtration fraction will be unchanged.

Concept Filtration fraction is the ratio of the GFR to the renal plasma flow. At rest, this has a value of about 0.2. This means that a fifth of the total renal plasma flow is filtered. If the renal sympathetic nerve vasoconstricts only the preglomerular afferent arteriole (which never happens in vivo, see next question 3), we can think about the effects on RBF and the GFR.

There will obviously be a decrease in RBF (the renal plasma flow is just ~55 % of RBF if the hematocrit is 45 %) with the increased vascular resistance. The effect on GFR will be mediated by any effects of sympathetic nerve on the Starling's forces that contribute to the net filtration pressure that produce the GFR. Since the glomerulus is "downstream" from the afferent preglomerular arteriole, the hydrostatic pressure that promotes filtration will be reduced. GFR will be decreased.

Since sympathetic action decreases both the renal plasma flow and the GFR, the filtration fraction is unchanged.

This is a good place to talk a little more of the cause and effect mechanisms in explaining physiology. In the above scenario, the renal sympathetic nerve activity is the initiating cause acting on the afferent arteriole. The net effect is an unchanged filtration fraction since both GFR and RBF is decreased in parallel.

If the initiating cause is stated as a change in the filtration fraction (FF), let us say a increased FF, then the mean glomerular oncotic pressure will be higher. The student could then reason that an increased mean #gc would lead to a reduced net filtration pressure and hence a decreased GFR. Thus, if we compare the two cases, the former has a reduced GFR/unchanged FF (because RBF also decreases) and the latter, an increased FF/reduced GFR.

If the student is discerning, it will be noted that the former is more physiologic. This is because an initiating cause given as an increased FF could already be due to a greater GFR. So, it becomes a case of circular thinking to work out how this higher FF will effect GFR!

3. What is the expected effect of sympathetic action on Pgc and GFR?

Answer The renal sympathetic nerve will decrease the GFR and the effect is due to a reduced RBF and not through any predictable change in glomerular hydrostatic pressure.

Concept It ought to be stressed to students that both the renal arterioles are innervated by renal sympathetic fibers. In other words, sympathetic vasoconstriction or decreased sympathetic vasodilation will occur concurrently on both the afferent and efferent arterioles.

The glomerulus is "downstream" from the afferent arteriole and "upstream" from the efferent arteriole (imagine the renal circulation as a "bloody" river; bloody used here not as a swear word but as an adjective!). As such, afferent vasoconstriction will lower the Pgc and the vasoconstricted efferent arteriole will heighten the Pgc. Thus, it is not easy to predict the overall effects of renal sympathetic nerve on the Pgc.

However, the renal sympathetic action will always reduce the RBF since vasoconstriction of either afferent or efferent will decrease renal perfusion. Whenever RBF changes, there will be a parallel change in the GFR in the same direction. The student might be surprised to learn that this effect of RBF on GFR is best explained, not by any predictable effects on Pgc as stated above, but by the inverse changes in the mean glomerular oncotic pressure #gc when the RBF changes. Looking back at the GFR formula, the net filtration pressure is (Pgc minus mean #gc minus Bowman capsular pressure), increased RBF will result in a decreased mean #gc giving an increased GFR.

The way to comprehend this RBF/#gc phenomenon is to imagine that the rise in the #gc along the glomerulus takes a comparatively longer time when the renal perfusion increases.

This also means that the calculated net filtration pressure, assuming it approaches 0 mmHg will be reached at a point nearer the efferent end of the glomerulus. We can view this as a larger capillary area where net filtration occurs. The GFR is higher with a greater RBF because increased RBF causes a lower mean #gc.

4. What is the role, if any, of sympathetic nerve on renal autoregulation?

Answer The extrinsic sympathetic innervations to the renal arterioles are not a contributing input to the Intrinsic renal autoregulation mechanism.

Concept By the term "intrinsic", renal autoregulation is able to maintain a constant RBF over a defined range of blood pressure fluctuations, independent of extrinsic nerve or circulating hormonal actions.

However, if the conditions in the body require a priority in a dominant renal sympathetic nerve activity, the neural input will override or "masked" the underlying RBF autoregulatory mechanisms.

To illustrate, the graphical description of renal autoregulation shows a controlled flow plateau over the blood pressure changes from 60 to 160 mmHg. This autoregulation was observed in an in vitro laboratory setup where only the intrinsic renal mechanisms that maintain flow were documented.

When the blood pressure drops in the body to 80 mmHg, would renal autoregulation still be effective? In this situation, the dominant renal sympathetic action is more essential to normalize the arterial blood pressure. The vasoconstriction of renal arterioles is part of the baroreflex sympathetic compensatory increase in total peripheral resistance. During hypotension, the autoregulatory mechanism should vasodilate the afferent arteriole in order to normalize the RBF. However, the transient reduction in RBF by renal sympathetic nerve pre-dominates and the intrinsic autoregulation is "masked."

This physiologic weightage on renal sympathetic nerve is also seen during exercise. There is some increase in arterial blood pressure during physical activity. The RBF, however, is not effectively autoregulated to remain unchanged. There is a need to redistribute the cardiac output to provide more perfusion to the skeletal muscles during exercise, the blood vessels in the muscle experiencing vasodilatation.

At the renal vasculature, the increased sympathetic action vasoconstricts the arterioles. Blood flow to the kidneys is relatively reduced to channel more of the greater exercise cardiac output to the muscles. This renal vasoconstriction also serves to maintain the blood pressure during exercise (since the arterioles in the muscle tissues are dilated, and there is an overall drop in total peripheral resistance).

5. How does a high protein diet affect, if any, the RBF?

Answer A high protein diet will lead to an increase in RBF and hence the GFR.

Concept A high protein diet does not postprandially change the plasma oncotic pressure. The plasma oncotic pressure is due to the plasma proteins, and this osmotic pressure is a major force that affects the capillary dynamics at the microcirculation.

Proteins are digested and the component amino acids are absorbed into the blood from the intestinal lumen. The plasma amino acids are higher during and after a meal.

Students who think that a high protein meal increases the plasma oncotic pressure (#c) will reason that the GFR will decrease since the net filtration pressure will be lower when the #c (#gc) is higher.

The GFR actually increases within an hour after consuming a high protein dinner (all you can eat steak buffet!). The hyperamino acidemia is the reason for the increased GFR. The mechanism interestingly involves the macula densa tubule-glomerular feedback.

The increased filtered amino acid load leads to a reduction in the NaCl sensing by the distal tubular macula densa. This is because at the proximal tubule, the higher filtered amino acids will promote sodium reabsorption via the sodium-coupled, secondary active reabsorption of amino acids by the proximal epithelial cells.

The macula densa (McD) detects less of the electrolyte in the distal tubular fluid and "assumes" that this could be due to a decreased RBF/GFR. The McD responds by transmitting paracrine signals to the afferent arteriole (less vasoconstrictor and/or theoretically, more paracrine vasodilators). The McD activates this effect of the intrinsic renal autoregulatory mechanism and the RBF increases as the afferent arteriole vasodilates.

6. How is the renal handling of inulin used to derive the value of GFR?

Answer The GFR is derived from using the unique renal handling of inulin, where filtered inulin load is equal to the excreted inulin load.

Concept Almost all known solutes are processed by the kidneys in at least two ways—filtration and reabsorption (or secretion). Many are filtered, reabsorbed and secreted, e.g., potassium cations.

The plant molecule inulin is unusual in that all the filtered load of inulin is excreted and in addition since there is no tubular secretion or reabsorption of inulin; filtered inulin load = excreted inulin load.

Putting in the components of this relationship,

GFR × plasma concentration of inulin

= V, urine flow rate × urine concentration of inulin.

Re-arranging, GFR becomes = excreted inulin load/plasma

concentration of inulin or Uin × V/Pin

Looking at this ratio, Uin × V/Pin is the value of an imaginary volume of plasma that has been "cleared" of inulin per time, and this "cleared" inulin then appears in the excreted load in urine.

Since inulin is freely filtered, and enters the Bowman's capsule easily with the filtered water, the GFR is equal to the volume of plasma water "cleared" of inulin/time, and this same value of plasma fluid filtered/time enters the Bowman's capsule.

Therefore, for any solute that is filtered and reabsorbed back into the circulating plasma, the renal clearance will be less than the clearance for inulin. For solutes, in particular, organic solutes/metabolites that are filtered and secreted before excretion into urine, their renal clearance will be more than inulin clearance.

For solutes that are reabsorbed and secreted, the net secretion or net reabsorption of the solute will determine the value of their renal clearances in comparison with inulin.

Before the renal clearance was conceptualized and the unique handling of inulin was found, there was a suggestion that the tubules secrete urine. This is now seen as incorrect and obsolete. There is no secretion of water by the nephrons. Remember that the daily GFR is an extremely large volume at 180 L. On average, we were about 2 L of urine per day depending whether you are in tropical Malaysia in December or in cold Scandinavia. Thus, there is no necessity for tubular secretion of water. The renal handling of water is just

Excreted = filtered + reabsorbed.

7. How do the peritubular capillary Starling's forces compare with the forces at the glomerulus?

Answer The plasma oncotic pressure is higher than the hydrostatic pressure along the length of the peritubular capillary.

Concept In the glomerular capillary, the hydrostatic pressure starts high at ~50 mmHg and is relatively stable along the glomerulus, sustained by the efferent high resistance smooth muscle structure. The glomerular oncotic pressure, lower than the glomerulus along the capillary however rises to about 40 mmHg due to the high filtration fraction.

The peritubular blood is the end glomerular blood that exits from the efferent arteriole. Thus, the peritubular blood has an elevated oncotic pressure compared to renal arterial blood that supplies the glomerulus. The high vascular resistance of the postglomerular efferent arteriole causes a significant drop in the hydrostatic pressure in the peritubular capillary, to less than 20 mmHg. We have a capillary network

Renal handling of solutes by the nephron filter reabsorb and/or secrete the solutes. The excreted amount of solute, E, is then dependent on either F − R, F + S, or F − R + S

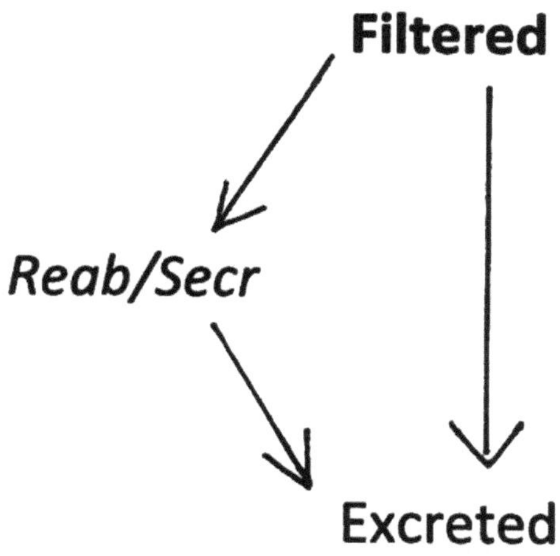

that supplies the renal tubules in which the plasma oncotic pressure is greater than the hydrostatic pressure along its entire length. This will generate a net reabsorptive Starling's force at the peritubular capillary. This unique capillary dynamic, the student will appreciate is nicely tuned to the functions of the renal tubules in reabsorption of water, electrolytes, and solutes.

The inquiring students may ask "What about tubular secretion?" For secretion, the solutes are generally organic compounds or metabolites. These solutes are transported bound to plasma proteins. The free solute is filtered, and the tubules also secrete the organic solute. There is an equilibrium between the free and bound solute, so secretion from the peritubular capillary will still occur. For active transport, there are organic cation transporters (OCT) and organic anion transporters (OAT) at the baso-lateral membrane of the proximal tubules. The passive secretion will depend on the availability of a concentration gradient between tubular fluid and the peritubular capillary/interstitium; there is no requirement for membrane transporters if the organic solutes can transverse the cell membranes down its concentration gradient.

8. How do intrarenal prostaglandins affect real blood flow?

Answer Intrarenal prostaglandins have vasodilatory action, and this effect serves to modulate and prevent an excessive constriction of renal arterioles.

Concept The renal sympathetic nerve vasoconstricts both renal arterioles. Circulating hormones like angiotensin II (AII) is a potent vasoconstrictor and enhances the effect of sympathetic action to increase the renovascular vascular resistance. AII is indirectly generated when the renal sympathetic nerve releases renin.

Concurrent with the action on renal arterioles, both the sympathetic activity and AII also increase the production and secretion of prostaglandin paracrines in the renal tissues. These prostaglandins relax vascular smooth muscle and counteract the vasoconstricting action of sympathetic nerve and AII. This intrarenal feedback

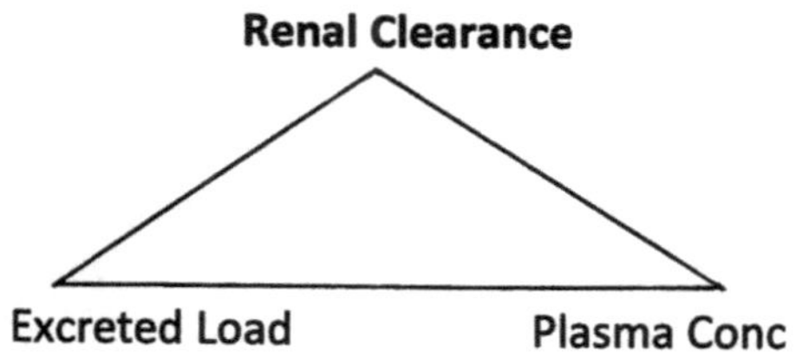

Renal clearance is the excreted load/rate (amount/min) divided by the plasma concentration (amount/vol) of the solute excreted. This gives a value that represents the "imaginary" volume of plasma that has been "cleared" of the solute that is found excreted into urine/unit time

provides some protection from potential renal ischemia when arteriolar constriction is intense.

Clinically, the action of renal prostaglandin vasodilators has implications for patients who are taking anti-inflammatory drugs, which inhibit prostaglandins. The kidneys in these persons would have reduced ischemic protection in situations when the renal arterioles constrict strongly.

9. Why is the renal clearance of creatinine suitable to monitor GFR in hospital setting?

Answer The renal clearance of creatinine approaches that of inulin clearance as the small amount of creatinine secreted and excreted is compensated by some laboratory false-positive for plasma creatinine.

Concept The renal clearance for inulin (filtered load = excreted load) is the definitive method for determining GFR. In the renal wards, it is not convenient to administer an exogenous solute like inulin to determine for changes in renal function. The renal clearance of an endogenous solute, creatinine is used regularly.

Creatinine is a metabolite, released into blood at a relatively constant rate. The excetion of creatinine is by filtration and secretion. Although the calculated renal clearance will be overestimated because of the excreted load Ucr.V, this is coincidently compensated by some overestimation of plasma creatinine by current laboratory analysis. Thus, the renal clearance of creatinine approaches that of inulin clearance.

In most cases, the physician requests for only one blood sample determination of plasma creatinine as an indicator of GFR. This is accepted, as there is an inverse relationship between plasma creatinine and GFR. Plasma creatinine (Pcr) will be elevated if GFR drops. The graph is not linear and the sensitivity is poor just below normal GFR value of 125 ml/min. However, the operating range in most clinical settings falls on the steel portion of the GFR/Pcr curve and is thus useful to monitor for improvement or deterioration in GFR as an indicator of renal function.

10. Which factor(s) in the GFR equation is altered by a vasodilator?

Answer The main change will be a decrease in the mean glomerular oncotic pressure which results in an increase in the net filtration pressure.

Humpty Dumpty sat on renal nephron wall

And he observe how the tubular fluid osmolarity fall

All the collecting ducts remains impermeable

No ADH, Polyuria, large volume is wee-ed again and again

The renal tubules handle solutes, and water filtered at the glomerulus. The membrane permeabilities of specific segments of the tubules are hormonally regulated with regard to transport of certain solutes and water. Adrenal aldosterone increases both luminal membrane permeabilities to potassium and sodium. Posterior pituitary vasopressin (ADH) increases the water permeability of collecting ducts during negative water balance

Concept This scenario reemphasizes the points made in question 3 above. Vasodilation occurs at the renal arterioles when the renal sympathetic nerve discharge is reduced. There is no dual innervation by parasympathetic vasodilator nerve to the renal arterioles (an important exception to the presence of parasympathetic vasodilator nerve is the fiber that regulates penal erection).

The RBF will increase, and this will increase the GFR. The higher GFR will also increase the filtered load (GFR × filtrate concentration of solute). There is a natural tendency to imagine that increased RBF should raise the hydrostatic pressure in the glomerulus as the primary change in increasing the net filtration pressure and GFR.

If we consider the separate effects of decreased sympathetic action on the afferent and efferent arteriole resistance, and how that changes the Pgc, it becomes clearer that Pgc is not the contributing factor.

Afferent vasodilation should increase the Pgc, while efferent vasodilation would permit more blood outflow from the glomerulus upstream and lower the Pgc. Thus, the opposite potential change in the Pgc with reduced afferent/efferent vascular resistance may leave the Pgc little altered and not necessarily an increased Pgc to produce a greater GFR.

Since the hydrostatic pressure in the Bowman's capsule is unaffected by increased RBF, the only Starling's force remaining is the (mean) glomerular oncotic pressure. We would, then, have to conclude that an increase in RBF with a vasodilator agent will lead to a decrease in the mean oncotic pressure in the glomerulus.

With a vasoconstrictor, the K_f component can also be reduced since the mesangial modified smooth muscle cells contract and decrease the surface area for filtration.

Chapter 12
Tubular Function

The renal tubules have many similarities to the digestive tract. This may sound odd. At both epithelial cells in the gastrointestinal tract and in the renal tubules, the reabsorption and secretion processes are key events. The epithelial cells are polarized cells and serve the unidirectional reabsorptive and secretory transmembrane transport events.

The term "Renal handling" refers to the triad renal processes of filtration, reabsorption, and secretion. What is excreted in urine is the net event of the three renal epithelial cellular activities. The general sequence is Excreted = Filtered—Reabsorbed + Secreted. Not all solutes handled by the nephron undergoes all three processes. Glucose is filtered and reabsorbed—there is no tubular secretion of glucose. This is the same nephronic scenario for water and sodium. The secretion of water to produce urine is a nineteenth-century theory!

The filtrate has a long journey from the Bowman's capsule to the collecting ducts, en route the loop of Henle (the original U-tube!). Some solutes like amino acids and glucose is transported entirely by the proximal tubules. More commonly, different segments of the nephron and collecting ducts modify the solute concentration in the tubular fluid, either by secretion or reabsorption or both.

As in the intestines, water reabsorption flows solute absorption. This is named "iso-osmotic" water reabsorption at the proximal tubule. The glucose in chyme and glucose in filtrate are transported similarly by secondary active mechanism using sodium-linked glucose cotransporter energized by the Na/K ATPase. Both the renal epithelial cells and the gastric parietal cells secrete hydrogen ions, for blood pH control and gastric digestion, respectively. The generation of hydrogen ions in renal tubular and gastric parietal cells require the catalytic action of carbonic anhydrase.

For organic solutes (other than especially glucose, amino acids obviously), being products of metabolism of foods or drugs, they are in general filtered and secreted. There are family groups of luminal/basolateral transporters that are dedicated to the active secretion of either organic cations or organic anions. The cationic and anionic polyspecificities of these organic solute transporters make economic sense when we consider the thousand of organic molecules that need to be secreted by the renal tubules.

H. M. Cheng, *Physiology Question-Based Learning*, DOI 10.1007/978-3-319-12790-3_12

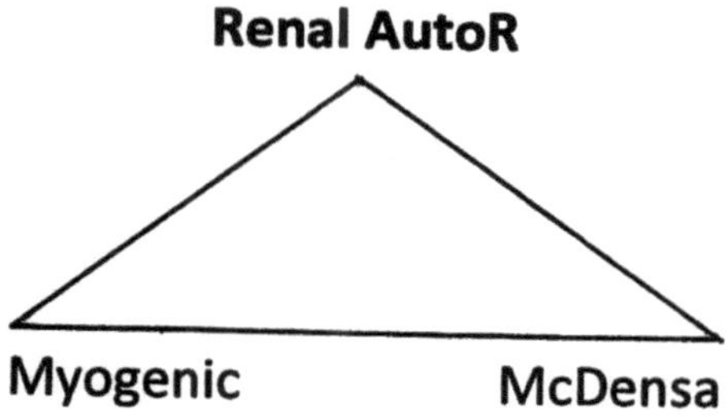

Two intrinsic mechanisms effect renal autoregulation of blood flow and glomerular filtration rate. The direct preglomerular afferent arteriolar response is called the myogenic mechanism. The *macula densa* of the distal tubule is involved in what is known as the tubuloglomerular feedback

Urea, produce of protein metabolism is a little unusual in being reabsorbed at the proximal tubule and secreted at the loop of Henle during urea recycling (reabsorbed from the inner medullary collecting ducts). This urea "merry-go-round" is potentiated by the hormone vasopressin and contributes to the hyperosmotic renal medulla that has an essential role in the urine-concentrating ability of the kidneys.

The epithelial cells of the renal tubules is a controlled long-processing line that determines the composition of the excreted urine. The urine chemical and osmotic profile is a reflection of the sequential cellular work that has gone on "upstream" along the tubule. The electrolyte balance, water balance, and pH balance are achieved by the reabsorptive and secretory capacities of the renal tubules. Trumpet the tubule!

1. Why does a decrease renal blood flow lead to uremia?

Answer There is less filtration of urea and more reabsorption of urea when the renal blood flow and thus, the glomerular filtration rate (GFR) decreases.

Concept Renal failure is accompanied by a reduction in GFR. The decrease in GFR can be due to intrarenal pathophysiology or to prerenal causes. The latter includes a drop in renal blood perfusion. In renal failure, there is an accumulation of solutes that are normally excreted in urine. This blood profile is called uremia ("urine in blood") and one major solute that is elevated in blood is urea.

Urea is handled by the nephrons via filtration and reabsorption at the proximal tubules (there is secretion of urea at the renal medulla that contributes to the hyperosmotic renal interstitium). When the GFR decreases in renal failure, less urea is filtered to be excreted. In addition, the proximal tubules also reabsorb more urea from the tubular fluid when the GFR is reduced.

The reason for this is that normally, the passive reabsorption of urea occurs when the urea becomes concentrated as water is reabsorbed at the early proximal segment.

When the GFR and thus the tubular fluid flow is reduced, the water reabsorption still takes place and the urea concentration becomes higher than usual due to the lower tubular fluid flow. More passive reabsorption of urea down a greater concentration gradient is promoted. Urea accumulates in the blood.

There is a proportionate relationship between urine flow rate and urea excretion. When the urine flow rate is reduced in renal failure due to poor GFR, urea excretion is decreased. Urea is raised in the blood. Note that when we talk of urea excretion, it is not urine urea concentration but urea excreted load. In diuresis, we can expect the urine urea concentration to be less, but the excreted urea load (concentration × urine flow rate = mg urea/time) is higher.

2. How is the sodium electrochemical gradient utilized for tubular transport?

Answer The sodium electrochemical gradient provides the potential energy for the solutes that are co- or countertransported when sodium moves down its gradient.

Concept Many solutes are transported by the renal tubules using secondary active transport. This secondary active mechanism utilizes in a majority of cases, a sodium electrochemical gradient to provide the potential energy for the solute transport. The sodium gradient is established and maintained by the Na/K ATPase pump that is active at the basolateral membrane of the tubular cells.

At the luminal membrane side, the filtered sodium concentration is the same as the plasma concentration at 140 mmol/L. The intracellular sodium concentration in the proximal cells is low ~15 mmol/L and the inside of the luminal membrane is slightly negatively charged as for all living cells. There is thus an electrochemical gradient of sodium and this cationic gradient is exploited by the renal tubular cells to move solutes.

Both glucose and amino acids in the glomerular filtrate are completely reabsorbed by the proximal tubular cells via sodium symporters. For tubular hydrogen ion secretion, a sodium-hydrogen exchanger at the luminal membrane is part of a similar secondary active transport for the proximal epithelial cells.

From the perspective of solute transport, we say that sodium helps or aids the movement of solutes across the luminal membrane. Viewed from the angle of sodium handling by the nephron, the movement of solutes linked to sodium gradient are actually moving or reabsorbing sodium into the tubular cells. Once inside the cells, the sodium is pumped out at the basolateral membrane by the Na/K ATPase into the interstitum and enters the peritubular capillary. The presence of different membrane transporters at the luminal/basolateral sides enables the functionally polarized renal epithelial tubular cells to transport solute unidirectionally.

3. How is potassium transported at the proximal tubule?

Answer Filtered potassium is passively reabsorbed at the proximal tubule via the paracellular route.

Concept At the proximal tubule, the reabsorption of potassium is similar to the renal handling of urea. The reabsorption is passive and requires prior generation of a potassium concentration gradient. The iso-osmotic water reabsorption at the early proximal tubule concentrates the tubular fluid potassium. The potassium then diffuses down its concentration gradient via the paracellular route to the interstitum to enter the peritubular capillary.

There should be no passive reabsorption of potassium transcellularly. The student can derive this claim by looking at some basic physiologic facts. Intracellular potassium is high in all cells, including at the proximal tubules. Filtered potassium concentration is the same as in plasma since potassium is freely filtered. Thus, the concentration gradient is uphill and very steep from the tubular fluid into the cells. It is not possible for potassium to be reabsorbed in to the proximal tubular cells. Even after being concentrated, the tubular fluid potassium will still be less than 10 mmol/L.

Can active reabsorption of potassium then take place instead? This is a possibility, if there exists a transport membrane pump for potassium at the luminal membrane. Na/K ATPase could be a candidate; however, all the Na/K ATPases are localized to the basolateral membrane.

4. Does the renal plasma threshold for glucose change in diabetes mellitus?

Answer The renal plasma threshold for glucose does not change in diabetes mellitus until renal complications begins to reduce the GFR in chronic diabetes.

Concept The threshold indicates that a new event will occur when the threshold is exceeded or when you walk through the threshold of the door, you enter into a new environment. The renal plasma threshold for glucose is a concentration threshold or limit. It is defined as the plasma glucose concentration above which, glucose begins to be excreted in to the urine.

The renal handling of glucose is filtration and tubular reabsorption. Filtered glucose is completely reabsorbed at the proximal tubule and no glucosuria is found in normal urine.

The tubular transport of glucose requires the sodium-linked glucose transporters (SGLT). These membrane transporters are limited and, thus, saturable if the filtered glucose load becomes high in the tubular fluid. The filtered glucose "load" is better understood as the rate of glucose filtration, given by GFR × Pg (plasma glucose concentration).

With increasing filtered load that follows increasing hyperglycemia, a point is reached when the filtered load just matches the completely saturated sodium-glucoase-linked transporter (SGLT, also called transport maximum rate for glucose, TmG). At this equilibrium, the elevated plasma concentration value is termed the "renal plasma threshold for glucose" (#Pg). Above #Pg, the filtered load exceeds the TmG and the unreabsorbed glucose is excreted in the urine.

The problem in diabetes mellitus is simply (a complex endocrine problem though!) uncontrolled hyperglycemia due to in adequate insulin action either reduced hormone secretion or poor receptor binding action at target cells. There is no change in the #Pg which is related to the proximal tubular function, TmG. The relationship is written as

$$\mathrm{GFR} \times \#\mathrm{Pg} = \mathrm{TmG}.$$

In a person with long-standing, chronic diabetes mellitus, the glomerular filtration is affected first before any further renal pathophysiology involving the tubules. Since the TmG is still normal, this means that a decreased GFR in chronic diabetes

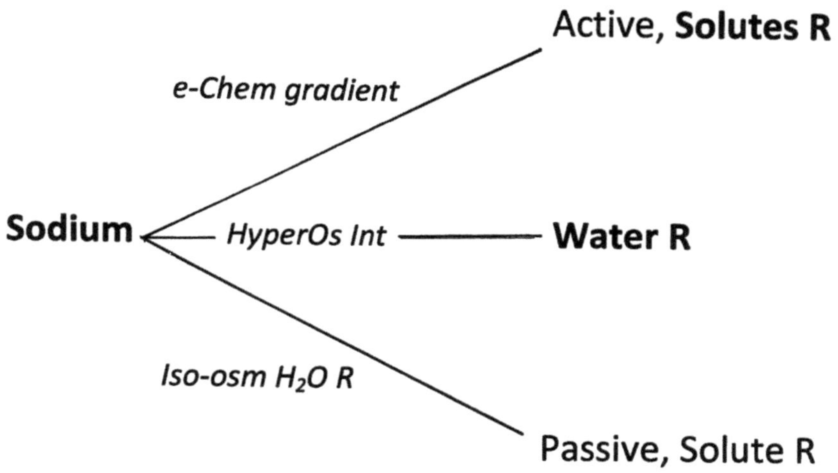

Sodium is central to nephron function. Reabsorption of many solutes is actively linked to sodium via secondary active transport, exploiting the sodium electrochemical gradient. The water reabsorption that follows sodium also concentrates several solutes for their eventual passive reabsorption (chloride, urea, K^+). Sodium is a key cation solute in the hyperosmotic renal medullary interstitium that drives H_2O reabsroption from the collecting ducts

will then be associated with a higher #Pg. Glucosuria will now occur at a higher plasma glucose concentration. The #Pg in a normal person is around 180 mg/dl. In chronic diabetes, glucosuria could still be absent even when the plasma glucose is already more than 200 mg%.

5. What is the role of glomerulo-tubular balance?

Answer The glomerulo-tubular balance (g-t) compensates for fluctuations in GFR and the filtered load by increasing or decreasing the proximal tubular reabsorption of water and sodium.

Concept Renal autoregulation of GFR is not a perfect intrinsic mechanism. Fluctuation in GFR still occur. A second line of defence to respond to acute changes in the GFR and, hence, filtered load is the g-t balance. The proximal tubule is the effector in this intrinsic renal pathway.

An increase in GFR will be compensated by an increase in proximal tubular reabsorption of fluid and solute, mainly sodium and vice versa. How does this glomerulus/proximal cross talk occur?

One explanation considers the peritubular capillary dynamics that change with fluctuations in GFR. As an example, an increased in GFR alone will increase the filtration fraction. This will lead to a higher end-glomerular oncotic pressure which will exit into and as the peritubular blood oncotic pressure. The higher GFR will decrease the hydrostatic pressure in the peritubular capillary. These changes in the peritubular Starling's forces combine then to promote more reabsorption of fluid and sodium.

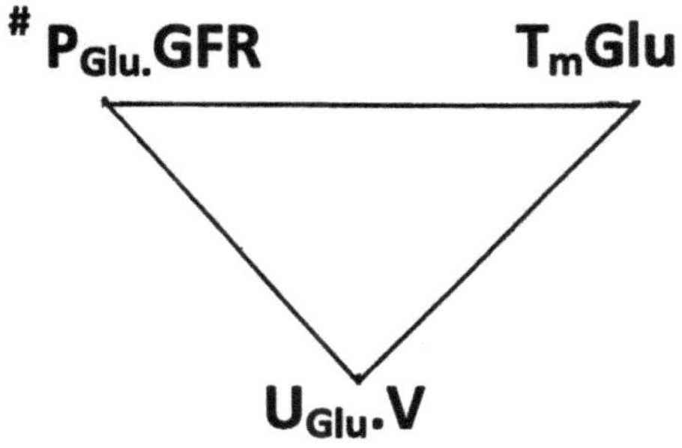

This renal *triangle* highlights the two parameters that determine if glucosuria will be present. The filtered load of glucose (amount/time) in a normal person, even during a high carbohydrate meal, will be less than the maximum rate of tubular reabsorption of glucose ($T_m Glu$). If the hyperglycemia exceeds the renal plasma threshold for glucose ($^{\#}P_{Glu}$), the proximal tubule can no longer recover all the filtered glucose. Glucose starts to appear in the urine (U_{Glu} > zero)

The student should note that this intrinsic g-t balance is effective in eu-volemic situation. If the extracellular fluid (ECF and, thus, the blood volume changes, priority will now be given to compensatory mechanisms that operate to normalize ECF/blood volume. This means that the g-t balance will be masked or overridden.

To give an example; when there is a contraction of ECF/blood volume, the appropriate compensations will include a reduced GFR. The lower GFR will help to conserve fluid. The proximal tubular reabsorption in g-t balance should also be reduced if the balance mechanisms is effective. However, in ECF contraction, the observed event at the proximal tubule is an increased reabsorption. Together the reduced GFR and the increased proximal recovery of fluid complement each other to preserve blood volume.

Likewise in ECF volume expansion, there will be an increased renal blood flow (RBF) and GFR and this is integrated with a decreased proximal tubular reabsorption in order to excrete more water and sodium. The priority laid on sodium and water balance (which involves extrinsic renal sympathetic nerve and hormone actions) overrides the intrinsic g-t balance mechanism.

6. How is glucosuria related to polyuria?

Answer The unreabsorbed glucose in the tubular fluid will interfere with the iso-osmotic reabsorption of water at the proximal tubule, and this results in an osmotic diuresis.

Concept Glucose filtered from the glomerular plasma is completely reabsorbed by the proximal tubule. The reabsorption of glucose (and other solutes) coupled to sodium transport generates a local osmotic gradient that then drives water movement at the proximal tubule, water reabsorbed both transcellularly and paracellularly.

Should the filtered load of glucose exceeds the transport maximum at the proximal tubule, the glucose remains in the tubular fluid and is osmoactive. The glucose left behind in the tubular fluid will be excreted in the final urine (glucosuria).

The presence of unreabsorbed glucose will reduce the iso-osmotic reabsorption of water, which is normally ~70% of the GFR. An osmotic effect producing what is termed osmotic diuresis.

This phenomenon is applied in prescription given to induce an osmotic diuresis. The solute mannitol is freely filtered but is not transported by any transporter at the luminal membrane of the proximal tubule. Mannitol then interferes osmoactively with the water reabsorption.

The carbonic anhydrase inhibitor when taken also causes a mild diuresis through the same osmotic effect. The enzyme inhibitor decreases the reabsorption of filtered bicarbonate at the proximal tubule which is normally ~80% of the filtered bicarbonate load. The increased bicarbonate anion in the tubular fluid is then osmoactive and exerts the mild diuretic effect.

7. Does the rate of filtration contribute to the excretion of hydrogen ions?

Answer The filtered hydrogen ion is an insignificant amount of total acid excreted daily since the normal plasma hydrogen ion concentration is around 40 nanoMole/L (10^{-9} Mole).

Concept The daily acid excreted is in the region of millimoles, about 70 mmoles. This represents the noncarbonic or nonvolatile fixed acid that has to be excreted by the kidneys. Renal failure is associated with a metabolic acidosis.

The free hydrogen ion in urine determines the pH of the urine. The minimum urine pH can decrease to at least 4.4. On the pH log scale, this represents a thousandfold higher concentration for H^+ in tubular fluid/urine compared to in plasma at pH 7.4. Most of the daily urinary acid load in mmoles is excreted as complexed urinary phosphate and ammonium ions.

The tubular secretion of hydrogen is obviously an active process against a limiting uphill concentration gradient of a thousand fold. Since even at pH 4.4, the free H^+ in urine is still of the micromole range. This highlights the importance of urinary phosphate and ammonium to enable the kidneys to excrete the daily millimolar amount of acids.

The renal epithelial cells of the tubule secrete H^+ to fulfill two cellular missions in acid base balance. The first is the use of secreted hydrogen to "fish" and reabsorb filtered bicarbonate. Around 80% of filtered HCO_3 anion is reabsorbed indirectly at the proximal tubule using this hydrogen "fishing" line dipped into the tubular fluid. The renal tubules also secrete hydrogen ion, generated from the hydration of CO_2 to carbonic acid. Newly synthesized bicarbonate ions from the dissociation of carbonic acid inside tubular cells are absorbed into the peritubular capillary.

The two membrane secretors of hydrogen ions are the sodium/H^+ antiporter at the proximal tubule luminal membrane and the H^+ ATPase at the intercalated cells of the collecting ducts.

8. How is the passive secretion of organic acids dependent on pH?

Answer For passive secretion of organic acids, a higher pH of the tubular fluid favours secretion and eventual excretion of the acid due to "diffusion trapping."

Concept Many organic solutes are either weak organic acids or bases. The proximal tubules actively secrete organic acids and bases. The organic solutes can also be passively secreted, if the downhill concentration gradient is available between the peritubular capillary and the tubular fluid.

To illustrate using organic acids, the unbound free acid (plasma protein binds to transport the lipophobic organic solute) is filtered. In the tubular fluid, there is some dissociation of the weak acid and the degree of hydrogen ion and organic anion separating is influenced by the pH of the tubular fluid.

If the tubular fluid becomes alkaline, more dissociation of the organic acid occurs. The charged organic anion species will increase. The organic anion will not be able to diffuse across the lipid rich luminal membrane and is "trapped" in the tubular fluid.

More dissociation will also mean that the concentration of the undissociated organic acid will be reduced in the tubular fluid. This will favour the provision of a concentration gradient for passive diffusion of the organic acid from peritubular fluid into the lumen of the renal tubules.

The overall process, where diffusion of the organic acid is enhanced and dissociation of the organic acid "traps" it in the lumen to be eventually excreted is called "diffusion trapping."

This tubular fluid pH dependence of passive organic acid secretion is utilized in the treatment of a person who has swallowed an excess of aspirin. Aspirin is salicylic acid. The basic bicarbonate is given to the overdosed patient to alkalinize the tubular fluid. This will promote passive secretion and more rapid elimination of elevated blood aspirin.

It should be noted that protein-bound organic solutes do not prevent them from being secreted by the tubules. This is because the protein-bound organic solute is in equilibrium with the free organic solute. As some of the organic solute diffuses, this is replaced by the release of some organic solute from the plasma protein carrier.

9. How does the renal clearance of urea compare to that of inulin?

Answer The renal clearance of urea is less than inulin clearance since there is net reabsorption of urea along the nephron.

Concept The clearance of inulin is used to determine GFR based on the fact that filtered inulin rate ("load") is equal to the excreted inulin load. The solute inulin is freely filtered and enters the Bowman capsule together in the filtrate.

The unit for renal clearance is volume/min and not the amount of solute/min. Clearance is a concept that imagines a certain volume of plasma "cleared" of a particular solute. Of course, in reality, there will be no portion of the circulating plasma that will be devoid of that solute.

For solutes that are reabsorbed after filtration, some of solute is returned to the plasma. In this case, the clearance of the solute into urine will be less than that for inulin, which is neither reabsorbed nor secreted.

For solutes that are both reabsorbed and secreted in different segments of the nephron, the renal clearance will depend on the net tubular process. For urea, ~50 % is reabsorbed at the proximal tubule. There is some secretion of urea at the loop of Henle of juxtamedullary nephrons during urea recycling, but this is the urea that is reabsorbed from the inner medullary collecting ducts.

Therefore for urea, there is still net tubular reabsorption. The renal clearance of urea will be less than the inulin clearance. More urea is excreted (mg/min) when the urine flow rate is higher. Urea clearance is higher with increasing urine volume excretion. If the clearance of urea is reduced, e.g., when GFR is low in renal dysfunction, the decreased urea clearance will result in increased plasma urea concentration seen in "uremia" of kidney failure.

10. How is the differential tubular reabsorption of sodium along the nephron linked to renal control of water balance?

Answer The proximal tubule reabsorbs water iso-osmotically following diverse sodium-solute transport. The reabsorption of sodium at the ascending loop of Henle contributes to the generation of hyperosmotic interstitium, essential for water reabsorption during negative water balance.

Concept The tubular handling of filtered sodium is linked to both water and sodium balance control. The major fraction of GFR, ~70%, is reabsorbed at the proximal tubule. This water reabsorption follows sodium-solute transport when a local osmotic gradient is generated. The iso-osmotic water reabsorption is termed obligatory and is not under direct hormonal or neural control (renal sympathetic nerve and angiotensin II does increase sodium reabsorption here).

The desending loop of Henle is basically impermeable to sodium and no sodium is reabsorbed. This results in progressive concentration of sodium in the tubular fluid in the juxtamedullary (jm) nephrons as water is reabsorbed, driven by the hyperosmotic renal medulla.

At the thick ascending loop of Henle, the sodium is actively reabsorbed and this active efflux of sodium helps to generate the stratified, hyperosmotic medullary interstitium. This segment of the jm nephrons is unusual in that it remains impermeable to water.

The hyperosmotic renal medulla is a prepared condition needed for reabsorption of water from the collecting ducts, downstream from the Henle's loop. During negative water balance, the hormone vasopressin (antidiuretic hormone, ADH) increases in plasma and the collecting ducts becomes permeable to water with the insertion of aquaporins at the luminal membrane of the ducts. Water then moves osmoactively since there is a ready hyperosmotic environment surrounding the medullary collecting ducts.

When sodium is reabsorbed at the principal cells of the collecting ducts under the stimulatory action of aldosterone, does water follow the sodium transport? The student can reason out that this "water follow sodium" phenomenon should take place, if there is simultaneous action of ADH. The ADH secretion is increased in negative water balance and the plasma aldosterone levels are higher in negative sodium balance. Are there situations, when there are both negative sodium and negative water balance? There are many common examples—post sweating, post blood donation. In fact any reduction in ECF volume would result in a negative sodium balance (all fluids contain sodium) and obviously reduced ECF water volume (an isotonic contraction, also a negative water balance is detected by volume/pressure receptors which then stimulate ADH via the hypovolemia). So, in these situations, we can say that water will follow sodium reabsorption at the collecting ducts, although this is not a necessary criterion at the jm nephrons, since the medullary interstitium is hyperosmotic and drives the water movement.

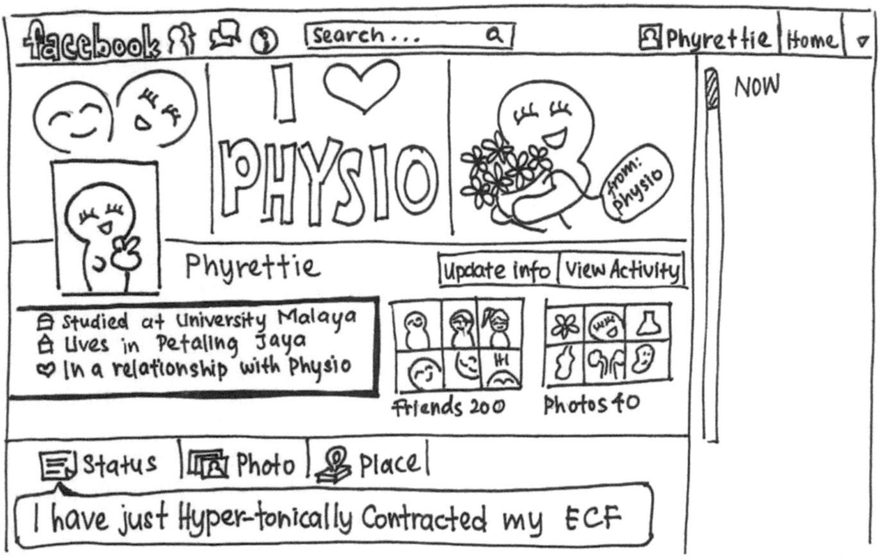

My student, Annabela Diong creatively drew this when I imagined a Physio female character called Phyrettie! There are six possible categories of extracellular fluid (ECF) changes based on volume and osmolarity alterations. During physical activity, sweating will hypertonically contract the ECF. In normal persons, only hypotonic contraction among the six ECF disturbances cannot be caused.... Why do you think that is the case??!!

Perhaps in the cortical nephrons where the interstitium is iso-osmotic to plasma at 300 mOsm/L, this “water follow sodium” event becomes significant, when the cortical nephrons conserve water.

A different scenario is a person who has a low sodium diet. This hyposodium intake will lead to a reduced ECF volume since total body sodium determines ECF volume. There is negative sodium balance and negative water balance, since the ECF water is reduced.

Chapter 13
Potassium and Calcium Balance

The two cations are present at relatively low concentrations in the extracellular fluid (ECF). Potassium concentration in blood is 4–5 mmol/L, and calcium is lower at around 2.5 mmol/L. The importance of maintaining these physiologic low concentrations of potassium and calcium imply that the homeostatic feedback control mechanisms for both cations must be sensitive and rapidly responding.

The physiologic control of potassium and calcium share the same design in having regulatory hormones that are not under the hypothalamo-pituitary axis (hpa) control. The secretion of steroid adrenal hormone aldosterone for potassium and parathyroid hormone (PTH)/vitamin D for calcium is not dependent on descending signals along the hpa axis.

The sensors for potassium and calcium reside on the cell membrane of their respective regulatory endocrine cells. K^+ sensors are on the zona glomerulosa cells in the adrenal cortex and Ca^{++} sensors on the parathyroid secreting cells.

Both ECF potassium and calcium affect nerve and muscle functions. The transmembrane potassium gradient determines the resting membrane potential of excitable cells. For calcium, changes in ECF calcium will alter the excitability of the nerve and muscle, probably via some steric action at the voltage-gated sodium channels.

The renal effector control of the electrolyte balance of potassium and calcium is via actions of the associated hormones acting on the tubular transport processes of these cations in the kidneys.

1. How is the filtered load for potassium and for calcium calculated?

Answer The filtered load of a solute is the product of the glomerular filtration rate (GFR) and the filterable portion of the solute; for potassium, this is the plasma concentration, and for calcium, it would be the free unbound calcium in plasma.

Concept The filtered load is actually the filtration rate of a solute. For potassium cation, the filtered load is the GFR multiplied by the plasma concentration (4–5 mmol/L) since potassium is freely filtered.

H. M. Cheng, *Physiology Question-Based Learning*, DOI 10.1007/978-3-319-12790-3_13

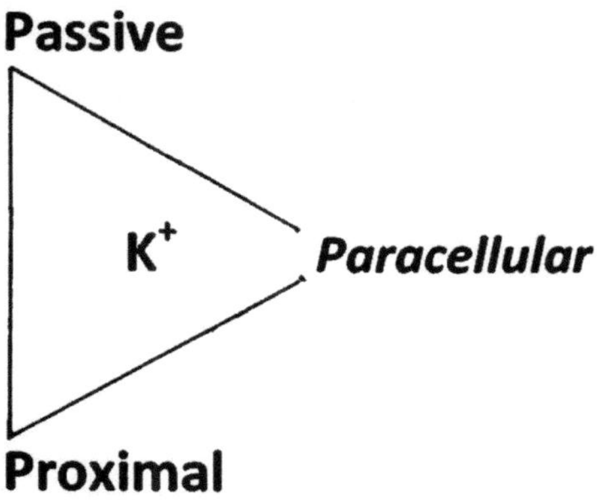

At least 70 % of the filtered potassium load is passively reabsorbed at the proximal tubule. Prior isoosmotic reabsorption of water at the early segment of the proximal tubule concentrates the tubular fluid K+. A chemical concentration gradient for K+ diffusion via the intercellular junction is generated. The absence of active K+ "pump" at the luminal membrane does not permit active transepithelial reabsorption of potassium

Calcium cations are 40 % bound by plasma proteins. The protein-bound calcium remains unfiltered in the glomerular capillary. The filtered load of calcium is the GFR multiplied by 0.6 (plasma calcium concentration, ~2.5 mmol/L). Of the 60 % filterable calcium in plasma, one fifth of the cation is also associated as complexes with anions including phosphate, sulphate and citrate. The calcium ionic complexes are also filtered. The free ionized plasma calcium is the biological active ionic species.

The student should note that after filtration, both the potassium and the calcium plasma concentrations at the end of the glomerulus have not changed.

2. How does pH of blood affect filtered calcium and potassium concentration?

Answer Decreased blood pH tends to increase plasma potassium as well as free plasma calcium and both cations will be filtered more at the glomerulus.

Concept The plasma potassium level is influenced by acid–base balance. Some acidosis tends to increase plasma hydrogen concentration. This operates as a result of intracellularly buffering in all cells that is then accompanied by a transmembrane K^+/H^+ exchange phenomenon in order to preserve body fluid electroneutrality in the ECF/intra cellular fluid (ICF) compartments.

For calcium, plasma albumin contains negatively charged sites that can bind either calcium or hydrogen ions. Acidotic plasma tends to compete and release more of the protein-bound calcium and a hypercalcemia can occur. Conversely, in alkalemia, the free, ionized calcium is decreased by more protein binding, leading to symptoms of hypocalcemia.

3. Are the respective cation sensors for potassium and calcium homeostasis located in the kidneys?

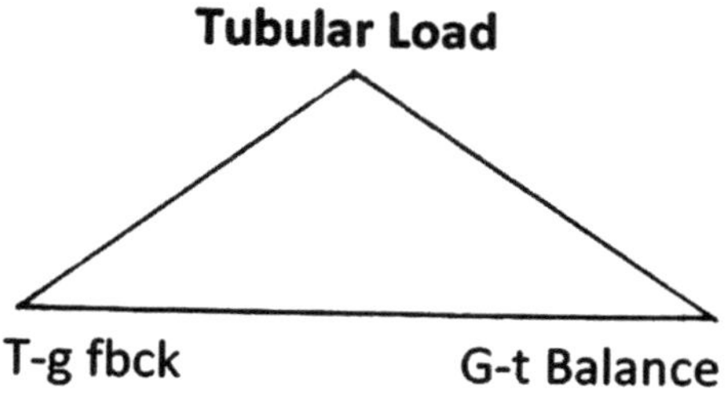

The maintenance of an optimal tubular fluid load from the Bowman's capsule to the rest of the long stretch of the nephron is provided by two intrinsic renal mechanisms. The first line of control is renal autoregulation which includes the myogenic and the tubulo-glomerulal feedback/macula densa responses. Since renal autoregulation of renal blood flow (RBF)/GFR is not perfect or fool-proof, a second line of regulation is the Glomerulo-tubular balance. This adjusts the degree of proximal tubular reabsorption of water and sodium in parallel with fluctuations in the filtered water and solute load

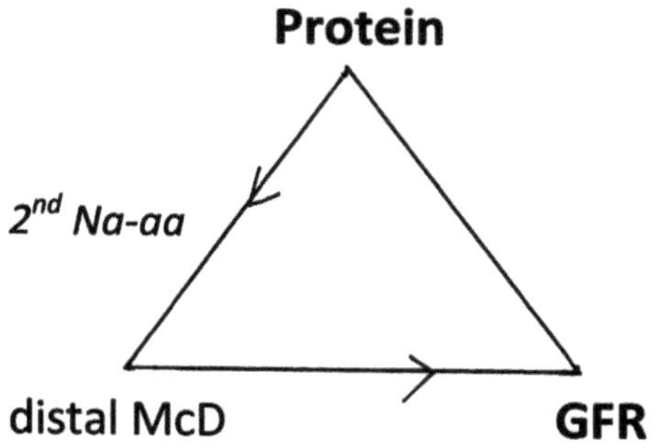

A high-protein diet leads to hyperamino acidemia. The greater filtered amino acid load will enhance sodium reabsorption via the secondary active Na-linked mechanism. The macula densa downstream at the distal tubule senses the reduced tubular fluid sodium/chloride. The paracrine effect from the McD on the afferent arteriole (both part of the juxtaglomerular apparatus) is to produce vasodilation and increased renal blood flow and glomerular filtration rate (*GFR*). This protein-*GFR* effect may be the rationale for a low-protein diet in patients with reduced renal function

Answer The potassium and calcium sensors are localized on the endocrine cells that secrete aldosterone and the PTH, respectively.

Concept Since the normal concentration of both cations are low in plasma, the control mechanisms that sense and monitor the cationic concentrations have to be rapidly responding. The natural site for receptors that detect changes in the calcium and potassium ECF concentrations would be the membrane of the respective endocrine cells.

For calcium parathyroid glands secrete the PTH in response to hypocalcemia. The PTH secreting cells have membrane sensors that serve this function. We would correctly infer that calcium membrane sensors are also found on endocrine cells that secrete calcitonin. Calcitonin reduces hypercalcemia and the calcium receptors respond to increased ECF calcium. If ECF calcium has a direct effect on synthesis of active vitamin D in renal cells, then these cells should be equipped with calcium receptor mechanisms that would be activated during hypocalcemia to increase vitamin D production. The triad of hormones, PTH, vitamin D, and calcitonin, all have actions on the nephrons of the kidneys to regulate calcium balance.

Potassium sensors that provide feedback in potassium balance/homeostasis are not located in the kidneys. The kidneys are the site for potassium reabsorption and hormonally regulated secretion. The K^+ sensors are membrane structures on cells in the zona glomerulosa of the adrenal cortex that secrete aldosterone. Aldosterone is a steroid hormone and secreted on demand, i.e., aldosterone is not made and packaged in vesicles that are released upon stimulation. There must be a signalling pathway from sensing hyperkalemia by these adrenal cortical cells to the release of aldosterone into the circulation.

4. How does the reabsorption of potassium at the loop of Henle affect calcium transport?

Answer The transepithelial potential generated subsequent to the activity of the Na/2Cl/K at the ascending loop of Henle favors calcium cation reabsorption.

Concept The triple cotransporter at the ascending loop of Henle (LoH) is by nature a neutral cotransporter, moving two cations and two anions simultaneously. However, some potassium diffuses back into the lumen. The result of the movement of potassium back across the luminal membrane leads to the Na/K/2Cl being an electrogenic transporter.

A lumen positive potential difference is created that helps to drive the transport of divalent cations like calcium and magnesium.

About 20–25 % of filtered calcium is reabsorbed at the ascending LoH via the paracellular route, driven by the electrical potential.

Since this triple transporter also reabsorbs sodium, the calcium reabsorption and the sodium reabsorption is coupled here. Clinically, this sodium/calcium linkage has important implications as loop diuretics acts on the triple symporter to inhibit sodium reabsorption. Thus, the loop diuretic also has secondary hypercalciuric effects. This action on increasing calcium excretion is used when prescribing loop diuretics for hypercalcemia.

5. How does renal failure affect plasma calcium?

Answer Renal failure causes hypocalcemia that then triggers a secondary hyperparathyroidism.

Concept There are a few reasons that explain the reduced plasma calcium in renal failure. The more obvious reason is that the source of active vitamin D is the kidneys. Under the action of the PTH, which has hypercalcemic effects, hydroxylation reactions that produce bioactive vitamin D is stimulated in the renal endocrine cells. If renal dysfunction affects these cells, plasma vitamin D may be insufficient to ensure adequate intestinal absorption of calcium.

Renal failure is commonly associated with a decreased excretion of urinary phosphate. Plasma phosphate accumulates. The hyperphosphatemia then leads to more complex formation with free ionized calcium. The latter is the bioactive free calcium. This decreases in plasma calcium, consequent from the increase in plasma phosphate, triggers secretion of the PTH.

The secondary hyper-PTH can be minimized by giving the patient "phosphate-binders." In normals, the PTH has a definite physiologic action in increasing phosphate excretion in urine by inhibiting phosphate reabsorption at the proximal tubule. The PTH increases distal tubular calcium reabsorption. These opposite renal actions of the PTH, combined with the increased bone resorption, stimulated by the PTH (that releases both calcium and phosphate into the blood) give a net increase in free ionized calcium that is produced by PTH action.

The phosphaturic action of the PTH should be remembered and it might help to think of the PTH as "PhosphaTuric Hormone"!

How might renal failure affect potassium balance. Normally more than 70 % of filtered potassium is reabsorbed at the proximal tubule. The fine-tuning of aldosterone regulation of potassium secretion takes place at the collecting ducts. Thus, the extent and site of renal damage will determine the overall effect on plasma potassium. An inadequate proximal reabsorption can potentially lose more potassium in the urine. On the other hand (on the other nephronic site!), a reduced ductal ability to secrete potassium in response to increased potassium dietary load can be hyperkalemic.

6. Does diuresis passively or actively affect potassium excretion?

Answer Increased tubular fluid flow enhances the gradient for passive diffusion of potassium from the principal cells into the lumen of the collecting ducts.

Concept Tubular transepithelial secretion of potassium is an active process. There are two steps in this active secretion. The basolateral membrane adenosin triphosphatase (ATPase) pumps potassium into the principal cells. Then, intracellular potassium diffuses passively at the luminal membrane down its concentration gradient in to the tubular fluid. The overall tubular secretion is active since the ATPase step is active.

The increased water excretion is accompanied by a greater secretion of potassium. This is effected at the luminal passive diffusive second step. Higher tubular fluid flow will tend to lower the tubular potassium concentration at luminal side of the principal cell. A steeper concentration gradient promotes more potassium secretion and hyperkaliuria results.

This hyperkaliuria occurs with action of loop diuretics. Interference of sodium reabsorption at the loop of Henle causes an osmotic effect due to the increased tubular fluid sodium. Downstream at the collecting ducts, the osmotic diuresis secondarily causes more potassium loss into urine by secretion (besides the concentration gradient effect by greater tubular fluid flow, the greater sodium load also enhances sodium/potassium "exchange" movements in the principal epithelial cell and potassium secretion is increased).

A question that might be asked is about the action or rather "inaction" of antidiuretic hormone (ADH) during normal physiologic response to positive water balance. Does the water diuresis in the absence of ADH also result in some kaliuric effects? That would not be a physiologic side complications in electrolyte control. ADH has been shown to have a stimulatory action on tubular potassium secretion. Thus, the suppression of ADH reduces somewhat potassium secretion and this counteracts the kaliuric effect of diuresis.

7. Would you expect the renal compensation for negative sodium balance to result in a secondary hypokalemia?

Answer Potentially, the increased aldosterone action to compensate by stimulating renal sodium reabsorption could also increase potassium secretion, but this will not make sense or be physiologic.

Concept Both the homeostasis of ECF sodium and potassium require the key regulatory hormone aldostereone. The target cell for aldosterone action in sodium and potassium control is also the same principal cell of the collecting ducts. Aldosterone acts by increasing the activity of the common membrane pump shared by sodium and potassium, Na/K ATPase at the basolateral side of the principal epithelial cell. At the luminal membrane, aldosterone acts to increase the membrane permeability to sodium and potassium by the addition of more sodium and potassium ion channels respectively.

Thus potentially, a compensation and normalization in response to a disturbance in one cation could lead to a secondary imbalance in the other cation.

However, in negative sodium balance, the ECF and hence blood volume is reduced. This means the renal blood flow would be decreased and the GFR is then less than normal. The decreased tubular fluid flow, "downstream" at the collecting ducts, as a result of the reduced starting GFR would affect potassium secretion. Lower tubular fluid flow will tend to slow potassium entry across the luminal membrane into the lumen.

This fluid effect to decrease diffusion of potassium provides the counterbalance to any secondary effect of aldosterone (induced by the negative sodium balance) to stimulate potassium secretion. The potassium balance is unaltered, which vindicates the physiologic design of the human body.

8. Compare how renal compensate for plasma potassium changes in diarrhea and in vomiting?

Answer Potassium is lost in both vomitus and fecal water. The aldosterone secretion would be inhibited to reduce renal excretion of potassium.

Concept Hypokalemia is a problem with vomiting and diarrhea. In addition, the volume contraction will tend to sustain the hypokalemia due to the ECF/blood volume contraction. Decreased vascular volume will trigger the activation of the sodium conserving renin-angiotensin-aldosterone system. Aldosterone stimulates ductal potassium secretion. There is thus the balance between the direct effects of hypokalemia in suppressing adrenal aldosterone secretion and the indirect effects of hypovolemia-induced aldosterone release.

In the metabolic alkalosis resulting from vomiting, the alkalosis will lead to the transmembrane hydrogen/potassium exchange activity as part of the general

Integrative Physiology

the Kidneys and the Heart in integrated blood pressure control

BP = CO x TPR

The kidneys are partners with the heart in maintaining ECF/blood volume. The blood volume is a major determinant of blood pressure, and this fact accounts for the pivotal role of the kidneys in what is termed the "long-term" control of blood pressure. The kidneys' volume regulation is tied to the cardiac output determinant of blood pressure

cellular buffering mechanism. This H/K interchange during alkalosis shifts potassium into cells and aggravates the hypokalemia.

There are other potential inputs that affect ECF potassium during hypovolemia. Volume/pressure sensing will trigger the effector sympathetic nerve activity. The adrenal catecholamines are secreted by the sole controlling neural action of sympathetic cholinergic nerve. Circulating adrenaline has negative or positive actions on the cellular uptake of potassium which depends on whether the alpha or beta adrenergic receptors are bound respectively by the hormone.

9. How is energy used, directly or indirectly to power potassium transport along the nephron, at the proximal tubule, loop of Henle and the collecting ducts?

Answer The sodium/potassium ATPase is the unifying energy nucleus that sets up conditions for potassium reabsorption and secretion along the nephron. Starting at the collecting ducts and going upstream, we first see the basolateral membrane ATPase pumping potassium into the principal cells of the ducts. Potassium then diffuses into the tubular fluid down its concentration gradient. Here, the ATPase maintains the high intracellular potassium concentration for the downhill diffusion of potassium into the lumen.

Midstream at the ascending loop of Henle, we can consider the reabsorption of potassium as a sodium-linked secondary active transport. The ever-faithful ATPase maintains a low intracellular sodium concentration in the "loopy" cells. There is then a sodium electrochemical gradient across the epithelial cells. The potential energy in this sodium gradient is then exploited at the luminal membrane by the triple Na/2Cl/K cotransporter. Potassium is brought into the cells and will presumably diffuse out at the basolateral membrane. Some potassium will also leak back into the lumen and this generates the lumen positive potential that is utilized for reabsorption of the cations, calcium and magnesium paracellularly.

At the proximal tubule, the energy expended by the Na/K ATPase reabsorbs ~70 % of the filtered sodium, mainly via solute-coupled sodium mechanisms. Many of these are sodium symporters, but the sodium-hydrogen antiporter should not be forgotten as this proximal membrane transporter is the major renal tubular secretor of hydrogen ions.

Water is then reabsorbed isoosmotically following the sodium/solute reabsorption. Upon water reabsorption, the potassium is concentrated in the tubular fluid and at the late proximal segment a concentration gradient for potassium is present between tubular fluid potassium and interstitial fluid potassium. Potassium is reabsorbed passively via the paracellular route (tight junctions are not tight enough to prevent K^+ passage!). At least 70% of filtered potassium is recovered into the circulation at the proximal tubule in like fashion.

This is an indirect, multistep mechanism from the ATPase-driven active sodium transport to water movement to generation of K^+ gradient for passive reabsorption.

10. How does the renal handling of potassium change in a low potassium diet?

Answer On a low potassium diet, the collecting ducts can begin to reabsorb potassium instead of secreting the cation when on a normal diet.

Concept Potassium concentration in the ECF is kept at a low level of less than 5 mmol/L. Each day, on a normal diet, excess potassium is added to the ECF. The homeostasis of potassium balance by the kidneys include hormonally fine-tuning the degree of potassium secretion at the principal cells of the collecting ducts. The adrenal corticosteroid hormone, aldosterone fulfills this function.

The intercalated cells of the collecting ducts can reabsorb potassium. This takes place if there is the need to maintain plasma potassium during an episode of low potassium dietary intake. The luminal membrane of the intercalated cells have the transporter, potassium/hydrogen ATPase exchanger. This is similar to the H/K ATPase that secretes acid at the gastric parietal cells.

Potassium is actively pumped into the cells from the tubular fluid against its concentration gradient. From inside the intercalated cells, potassium then diffuses out at the basolateral membrane into the interstitium and then into the peritubular capillary.

Aldosterone has also a positive action in the secretion of hydrogen ions by stimulating an H^+ ATPase also present on the luminal surface of the intercalated cells. The other hydrogen membrane pump, H/K ATPase is not affected by aldosterone as this would mean that aldosterone also increases potassium reabsorption (which will oppose its stimulation of potassium secretion at the principal cells).

Hypokalemia stimulates tubular ammonia synthesis and increases the ability of the nephron to excrete hydrogen ions as ammonium NH_4^+. In hypokalemia, potassium exits the renal cells, and there is a mutual replacement of the cation with hydrogen ion. The resulting decrease in intracellular pH stimulates NH_3 synthesis from glutamine.

Chapter 14
Water Balance

Our kidneys excrete urine of varying volume—dilute and concentrated urines. The excreted solutes must be dissolved in water (we obviously cannot urinate solids!). The range of the volume and osmotic concentration of urine reflects the ability of the kidneys to regulate and maintain a constant extracellular fluid (ECF) osmolarity. Changes in ECF are rectified by the kidneys and translated to changes in urine volume/osmolarity. Osmoregulation is the same as regulating water balance. Since ECF sodium concentration is the main determinant of ECF osmolarity, we have three controlled parameters that refer to the same physiologic homeostasis. These are water balance, osmoregulation, and sodium concentration (like a three-in-two coffee pack!). We can quite correctly consider the hypothalamic osmoreceptors as receptors that monitor ECF sodium concentration.

If we are overhydrated, we use the term positive water balance. A positive water balance will at the same time lower the ECF sodium concentration (hyponatremia) and osmolarity (hypoosmotic). The student should note that drinking a large volume of water decreases the sodium *concentration,* but the total body sodium (sodium balance) is unchanged [see next set of questions on sodium balance; receptors that monitor sodium balance are a family of volume/pressure receptors].

The kidneys are unusual in having a medulla that is hyperosmotic. This is generated by a subpopulation of nephrons (juxtamedullary nephrons) and their associated peritubular vasa recta capillaries. The hyperosmotic medullary interstiium is essential for the kidneys to respond to negative water balance by reabsorbing more water, in concert with the action of the hormone vasopressin (antidiuretic hormone).

1. How are water balance and osmoregulation related?

Answer A positive water balance will lower ECF osmolarity and lead to excretion of a high volume, hypotonic urine.

Concept Most students are familiar with the book Thesaurus which list synonyms, words with related meaning. This provides the person a spectrum of words to describe the same phenomenon in different ways or from different perspectives.

H. M. Cheng, *Physiology Question-Based Learning,* DOI 10.1007/978-3-319-12790-3_14

Osmoregulation, control of water balance, or regulation of ECF sodium concentration all refer to the same physiological function

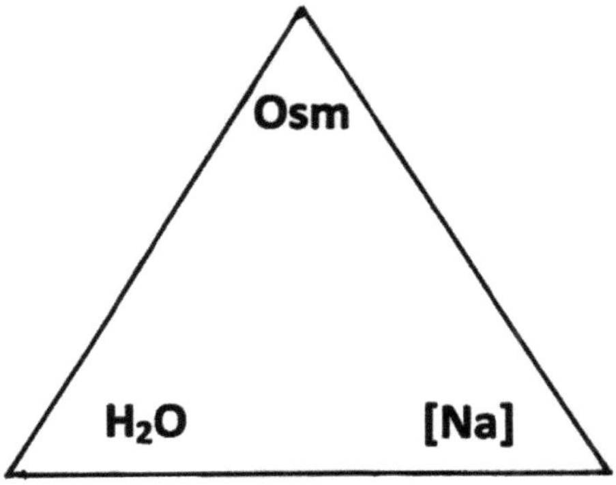

In physiology, the language and terminology commonly used also often include different names that describe the same physiological control events. Regulation of water balance and osmoregulation is one example.

Osmolarity of the ECF is determined predominantly by ECF sodium concentration. A change in the water balance is frequently the cause of either hypernatremia or hyponatremia. This association is important to note as most cases of clinical hyponatremia are not due to loss of sodium but retention of water (in Addison's disease, there is a loss of sodium).

A positive water balance occurs when a student in a physiology practical drinks a large volume of water. Potentially, this excess water will lead to hyponatremia and a hypoosmotic ECF.

The osmoreceptor sensing is rapid and within 30 min, the kidneys will excrete the excess water to increase and normalize the osmolarity. The water diuresis is effected as the antidiuretic hormone (ADH) secretion from the posterior pituitary is inhibited. The hyponatremia will quite soon return to eu-natremia.

The three physiological synonyms discussed here are osmoregulation, control of (ECF) water balance and control of ECF sodium concentration.

2. Why is there no need for water secretion from the renal tubules?

Answer The renal handling of water is simply filtration minus reabsorption since the glomerular filtration rate (GFR) is normally 99 % larger than the urine flow rate.

Concept The tubular secretion of water to produce urine is an obsolete nineteenth-century theory. With the use of inulin and the concept of renal clearance, normal GFR was found to be an immense large volume of plasma water filtered daily (180 L/day). In a 70 kg male adult with a plasma volume of 3 L, this is a sixty times total plasma filtration.

The tubular secretion of water would mean that water is transported from the peritubular capillary across the epithelial cells of the nephron into the lumen. With a GFR of 180 L/day and normal urine production of around 1 % of GFR (1.8 L/day), it is clear that no tubular secretion of water occurs in the kidneys (I tell my students that if they still write "water is secreted and excreted", their grades will be immediately secreted!).

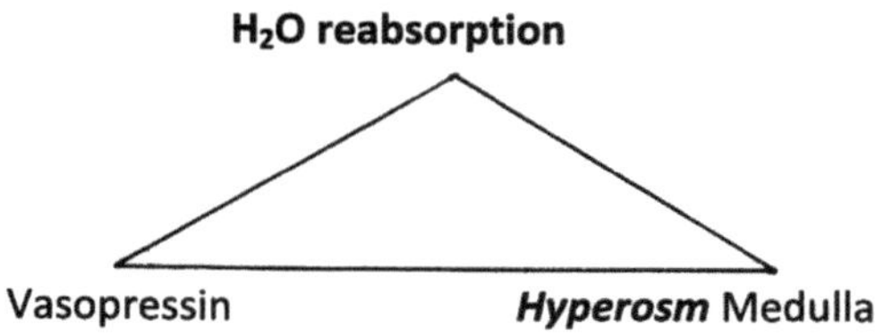

The reabsorption of water from the collecting ducts in the renal medulla requires the presence of two factors; a hyperosmotic medullary interstitium generated by the renal countercurrent mechanism and circulating vasopressin hormone from the posterior pituitary. In the renal cortex, only vasopressin alone is available. The water is still absorbed at the cortical collecting ducts but limited by the iso-osmotic cortical interstitium. Water from the hypotonic fluid that exits the loop of Henle will move out of the lumen when the collecting ducts become water permeable by vasopressin's action

There is polyruia in glucosuria of diabetes mellitus. The increased water excretion is due to the osmotic effect of unreabsorbed glucose at the proximal tubules. The osmotic diuretic effect in glucosuria does not "attract, pull" or drives water osmoactively into the lumen at the proximal tubules. In other words, it is incorrect to say that unreabsorbed glucose causes water secretion at the proximal tubule.

Rather, the excess glucose in the tubular fluid interferes and reduces the iso-osmotic reabsorption of water at the proximal tubules. Normally ~70 % of total glomerular filtrate is reabsorbed at the proximal tubules.

3. How do the water permeabilities of the loop of Henle in the renal medulla contribute to a hyperosmotic interstitium?

Answer The descending loop of Henle is permeable to water, the ascending loop is always impermeable to water, and the collecting ducts become permeable to water when the vasopressin acts.

Concept The juxtamedullary (jm) nephrons together with their associated peritubular vasa recta established and sustained a stratified hyperosmotic renal medulla. The jm nephrons generate the hyperosmotic medullary interstitium (hmi) and the vasa recta capillary blood flow preserves the hmi.

The ability of the jm nephrons and the vasa recta to create the hmi is due to the countercurrent flows in both the jm nephrons and the vasa recta. The tubular fluid flow in countercurrent, and the vasa recta blood flow is also countercurrent in the descending and ascending vasa recta. The student should not be misled by the visual 2-D representation in physiology texts that often give the impression that the countercurrent is between the tubular fluid flow in the nephron and the blood flow in the vasa recta. The overall jm nephron/vasa recta renal machinery that stratifies the medulla hyperosmotically is called the renal countercurrent mechanism.

The jm nephrons have unique membrane water permeabilities at different segments of the nephron. The descending fluid becomes increasingly concentrated because the water is reabsorbed, but sodium, the major tubular fluid solute is not reabsorbed. At the ascending loop of Henle (LoH), the epithelial cell is unusual is not allowing water to transverse it, transcellularly and even paracellularly. Sodium is reabsorbed here, actively at the thick segment of the ascending LoH, transported by the triple Na/K/2Cl membrane carrier.

The hyperosmotic fluid (~1300 mOsm/L) that enters the ascending loop, thus, becomes progressively diluted as it ascends and the fluid that exits the ascending Loh is always hypotonic (~100 mOsm/L). The combined effects of increasing sodium in the descending LoH and decreasing sodium in the ascending LoH with extrusion of sodium dynamically produce a renal medulla with increasing osmolarity, from 300 mOsm/L at the cortex to ~1300 mOsm/L in the inner medulla.

The fate of the remaining filtrate that travels down the collecting ducts will depend on the water balance in the body. In negative water balance, the ADH (vasopressin) is secreted from the posterior pituitary and acts to make the collecting ducts permeable to water. The availability of the hypertonic medullary interstitium then osmoactively drives the water reabsorption.

4. What role of the vasa recta maintains the hyperosmotic renal medullary interstitium?

Answer The countercurrent vasa recta acts as a passive exchanger and maintains the hyperosmotic interstitium by preserving the solutes in the medulla and removing reabsorbed water from the medulla.

Concept The juxtamedullary nephrons actively generate the hyperosmotic medullary interstitium (hmi). The peritubular vasa recta capillaries that course along the jm nephrons passively maintains the hmi. The countercurrent capillary blood flow is essential to enable the vasa recta to fulfill this function.

Imagine swimming inside and down the descending vasa recta. Since the vasa recta blood enters the renal medulla that is increasingly more hyperosmotic, the passive movement of solutes enter the vasa recta, and the water exits the vasa recta capillary into the interstitium. Thus, the blood at the tip of the vasa recta U-tube will also equilibrate with its surrounding and become ~1300 mOsm/L.

As the hyperosmotic blood ascends the vasa recta, imagine again swimming up and away into regions of the medulla with decreasing hyperosmolarity. The solutes that entered previously during the descending blood flow will now diffuse back into the medullary interstitium. Thus, little solutes are lost from the renal medulla, and the hmi is preserved. This passive movement and preservation of the hmi by the vasa recta is only possible, because the vasa recta is a countercurrent structure. Otherwise, there will be "washout" effect of the solutes by the descending capillary blood flow and the hmi cannot be sustained.

(The inquiring student may wonder at the exposure of the red cells to such hypertonic blood in the vasa recta).

For water, the water that exits the descending vasa recta is returned in to the ascending vasa recta as the capillary blood flows pass interstitium of decreasing hyperosmolarity. Again, we see that no water accumulates in the interstitium to disrupt the hmi. The student should also note that the 2-D diagrams in textbook may leave the impression that there are significant interstitial spaces between the ascending and descending limbs of the vasa recta (and also the ascending/descedign LoH). In reality, the two U-tubes (LoH and vasa recta) and the collecting ducts (CD) are packed closely together (visualize a cross section of the medulla as containing a closely associated group of five circles—two for LoH, two for vasa recta, and one for the CD).

The ascending vasa recta also removes water that is reabsorbed at other parts of the nephron. The descending LoH reabsorbs a smaller portion of the glomerular filtrate (~20%). The collecting ducts, under vasopressin action fine tunes the degree of water reabsorption in response to changes in water balance.

Therefore, the water reabsorbed from the descending LoH and the CD also enter the ascending vasa recta capillary and are carried away in the circulation. The hmi remains highly hyperosmotic.

5. What is the contribution of urea to water excretion?

Answer Urea contributes a major portion to the ability of the kidneys to produce the maximum concentrated urine of ~1300 mOsm/L.

Concept The highest osmolarity in the renal medulla is contributed by sodium chloride and also urea. Urea is recycled from the inner medullary collecting ducts, secreted into the LoH and circulates towards the CD again.

This recycling of urea between the CD and LoH occurs only when plasma ADH level is increased. This is because ADH also increases the permeability of the inner medullary CD to urea.

The hyperosmolarity in the medullary interstitium (hmi) will only be around 600–600 mOsm/L in the absence of urea recycling. In other words, during the negative water balance when ADH is secreted, the hmi will be maximum at ~1300 mOsm/L.

The cortical CD are made permeable to water, but not urea. So urea is concentrated along the CD. This allows urea to develop a concentration gradient at the inner medullary CD from where it recycles when the membrane becomes permeable to urea.

Conceptually, if the CD membrane is not made permeable to urea, the urea will be osmoactive in the lumen and this will oppose water reabsorption when the CD becomes permeable to water when ADH binds to the principal cells of the CD.

When secretion of ADH is inhibited during positive water balance, water excretion is increased with a higher urine flow. The urea excretion rate ($U_{urea} \times V$) is also increased during diuresis. During antidiuresis when water is reabsorbed at the CD, the urine is concentrated and the urea concentration in urine is also higher. But, the excreted urea load is lower than when the urine is dilute and hypotonic, but the urine flow rate is larger.

When there is no recycling of urea during increased water excretion in the biological absence of ADH, more urea is excreted in the urine.

6. What effects of a loop diuretic increase the water excretion?

Answer Loop diuretic causes increased water excretion by inhibition of sodium reabsorption at the LoH leading to a reduced osmotic gradient for water reabsorption.

Concept Loop diuretics inhibit the triple cotransporter Na/K/2Cl at the ascending LoH. This secondary active transporter reabsorbs sodium and is a major contributing mechanism that generates the hyperosmotically stratified renal medulla. The hypertonic medullary interstitium (hmi) is essential to enable the kidneys to concentrate urine during negative water balance.

By inhibiting loop transport of sodium, the hmi will be less hyperosmotic. More sodium remains in the tubular fluid and increases the fluid osmoalrity. The normal dilution of the fluid ascending the loop to produce "free water" no longer occurs. Downstream at the collecting ducts, water is reabsorbed when the ducts become permeable when acted upon by vasopressin secreted from the posterior pituitary.

The osmotic gradient between the increased fluid osmoalrity in the collecting ducts and the reduced hmi will be decreased. Even if the duct is permeable to water, less water will be reabsorbed. Water excretion as the final urine volume increases.

The effect of a loop diuretic can be said to cause an "osmotic dieresis," since it acts by osmotic interference.

Other categories of diuretics also inhibit sodium reabsorption and produce similar effects in increasing water excretion. The sodium/chloride cotranporter at the distal tubule is inhibited by the thiazide diuretics. The inhibitors of aldosterone action at the principal cells of the collecting ducts will also result in unreabsorbed sodium in the lumen. An osmotic diuretic action is effected.

With aldosterone antagonists, there is no secondary loss of potassium in the urine as occurs with loop diuretics. Action of loop diuretics alone leads to an increase in tubular fluid that arrives "downstream" at the collecting ducts. The principal cells respond to the increased sodium load by reabsorbing more sodium. Since aldosterone also acts to promote potassium secretion at the same target of principal cells, this results in hyperkaliuria and hence potential hypokalemia.

7. How do the osmolarity and volume changes affect ADH secretion after exercise loss of sweat?

Answer Both the hyperosmolarity and the hypovolemia stimulate ADH secretion via osmoreceptor and volume receptor pathway, respectively.

Concept Sweat is a hypotonic fluid. In fact, sweat is designed to be always hypotonic. Sweating results in a hypertonic contraction of the ECF. There is then a compensatory shift of fluid from the cells to the ECF. Imagine if sweat is hyperosmotic; the resulting hypotonic ECF will lead to more fluid shift into the cells and further contracts the ECF! Sweat is, thus, wonderfully hypotonic.

Sweating results in both a negative water and negative sodium balance (when body fluid is lost, there is always a negative sodium balance since all body fluid contain sodium). The osmoreceptor/ADH mechanism is most sensitive to ECF/plasma osmolarity changes. The hyperosmotic ECF will stimulate ADH secretion. The ADH increases water reabsorption at the kidneys to normalize the plasma osmolarity. Note that although the osmolarity is restored, the sodium balance is still negative, until the fluid loss itself is recovered.

The hypovolemia can also activate a reflex signal via the volume and baroreceptors to increase ADH secretion. Hypovolemia concurrent with hyperosmotic blood also increases the osm/ADH sensitivity.

As for ADH activation, the thirst neurons in the hypothalamus are also stimulated by hyperosmotic ECF and hypovolemia.

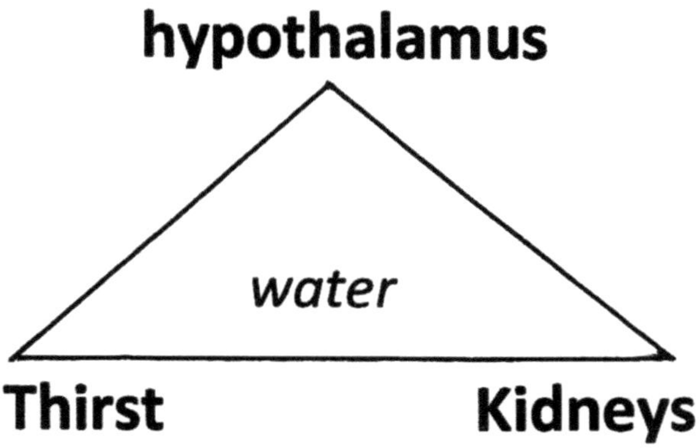

The brain is involved in maintaining our water balance or osmoregulation. The thirst center neurons are located in the hypothalamus. The hypothalamic osmoreceptors are associated with the posterior pituitary secretion of the hypothalamic neurohormone vasopressin that increases the water reabsorption in the kidneys. The micturition reflex for when we 'Wee' is coordinated by neurons in the brainstem and of course voluntarily by our cerebral cortex

The complete normalization of ECF/blood volume will require the body to respond to the negative sodium balance. This includes triggering sodium conservating mechanisms which involve renin-angiotensin-aldosterone pathway and increased renal sympathetic activity.

8. What is the relationship between osmolar clearance and free water clearance?

Answer Osmolar clearance is equal to the urine flow rate for urine that is iso-osmotic to plasma, and dilute urine will, thus, be the sum of osmolar clearance and free water clearance.

Concept The formula for osmolar clearance is the $U \times V/P$ ratio using osmotic concentration instead of specific solute concentration. The $U \times V$ will then be the excreted osmotic load and the P is the plasma osmolarity.

If the urine has the same osmolarity as plasma, the osmolar clearance will then be just V, which is the urine flow rate in ml/min.

We can imagine dilute urine as consisting of a portion of iso-osmotic urine and a portion of solute-free water (free water). Therefore, a dilute urine flow rate V becomes the sum of the osmolar clearance C_{osm} and the free water clearance, C_{water}.

Conversely, when concentrated urine is produced, the reduced urine flow can be viewed as the C_{osm} minus free water that has been reabsorbed. The latter is also termed negative free water clearance.

The ascending LoH of the jm nephrons (same in cortical nephrons?) reabsorbs sodium, but the membrane is impermeable to water (unique, unusual property, which implies no membrane aquaporins and very tight intercellular junctions). The tubular fluid that exits the ascending LoH is thus always hypotonic. The ascending LoH is described as the "diluting segment" and the reabsorption of sodium without accompanied water generates the free water. The fate of this free water at the collecting ducts will depend on the water balance in the body.

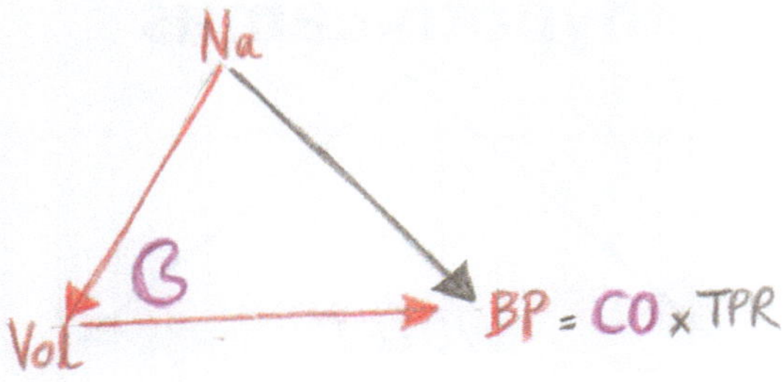

This triangle knowledge map links sodium balance with blood volume control by the kidneys. Note that this is total body sodium and not sodium concentration (vasopressin controls the ECF sodium concentration). Sodium balance is then physiologically associated with blood pressure. The renin-angiotensin-aldosterone system (RAAS) regulates sodium/volume balance, not sodium concentration. Sodium balance changes are thus monitored indirectly via volume/pressure vascular sensors

9. How does the ECF osmolarity and volume change in diabetes insipidus and syndrome of inappropriate ADH (SIADH)?

Answer In diabetes insipidus, the action of vasopressin (ADH) is reduced, resulting in high volume dilute urine leading potentially to a hypertonic contraction. In SIADH, uncontrolled, excessive ADH secretion leads to a hypotonic expansion.

Concept In physiologic homeostasis, the ECF osmolarity is controlled by feedback inputs from changing osmoalrity. Here, the fluctuation in osmoalrity is the cause and the compensation mechanisms involving ADH and the kidneys the effectors.

In clinical situations, an abnormal secretion or action of ADH becomes the primary cause, resulting in pathophysiologic effects on ECF volume and osmoalarity. In some lung tumors that secrete ADH (SIADH; syndrome of inappropriate ADH), there is an excessive water reabsorption from the nephrons independent of water balance in the body. There is retention of water and the ECF is enlarged in a hypotonic expansion. The hypotonic ECF (hyponatremia) can be life threatening as an excess water influx into neurons leads to brain swelling. Neurological symptoms ensue.

Conversely, a defective hypothalamo-posterior pituitary neural linkage will reduce the ADH secretion when needed. The kidneys cannot respond to negative water balance or increased ECF osmolarity although the hypothalamic osmoreceptors are stimulated. The kidneys continue to excrete a hypotonic large-volume urine. The hypertonic contraction of the ECF is not compensated for by the kidneys. However, a conscious person will have a greater thirst to drink as the hypertonic ECF will stimulate the thirst neurons also present in the hypothalamus. The plasm osmolarity in a diabetic insipidus person could, thus, be normal due to the increased water intake.

Note the increased water excretion in diabetes insipidus (DI) is due to little ADH action (can be receptor dysfunction in nephrogenic DI) while the polyuria in diabetes mellitus is produced by the osmotic diuresis that accompanies glucosuria. The dehydration in diabetes mellitus if the diuresis is high will stimulate ADH secretion since the hypothalamus is normal as pancreatic insulin is the problem.

10. How are the cortical nephrons involved in water balance?

Answer The cortical nephrons also reabsorb water as they share the same colllecting ducts with the juxtamedullary nephrons.

Concept The ability of the kidneys to produce concentrated urine is important during negative water balance, e.g., dehydration from sweating. The subpopulation of jm nephrons are dedicated to generating a hyperosmotic interstitium which enables the kidneys to concentrate urine during reabsorption of water.

The jm nephrons is less than 20 % of the total nephron population. The glomerular filtration rate is a combined value from every nephron in both kidneys. Thus, although the jm nephrons are the focus when describing the kidneys' unique ability to concentrate urine, the student should not forget that the crotical nephrons together contribute at least 80 % of the remaining filtrate that enters the collecting ducts.

If there is a positive water balance, the hypothalamic osmoreceptors are inhibited and ADH secretion is suppressed. ADH increases the permeability of the collecting ducts in both the cortex and medulla of the kidneys. In the absence of the ADH, the cortical and meduallry ducts are impermeable to water. The urine that exits the collecting ducts from the cortical and jm nephrons is hypotonic and large in volume.

In negative water balance, the cortical collecting ducts are permeable to water when acted by ADH. The tubular fluid is reabsorbed until osmotic equilibrium is reached at ~300 mOsm/L, the value of the osmoalrity in the cortical insterstitium. Remember that the fluid that flows from the LoH is always hypotonic due to the "diluting segment" of the ascending LoH that generates the "free water". Further water reabsorption from the collecting ducts proceeds at the renal medulla. The maximum hyperosmotic human urine is set by the highest strata of interstitial osmolarity in the inner medulla which is ~1300 mOsm/L.

A normal 1 % of GFR excreted as urine is 1.8 L/day. The urine volume can, thus, vary in response to water balance. The urine osmolarity, likewise, varies from a very dilute 50 mOsm/L to 1300 mOsm/L.

Chapter 15
Sodium Balance

The cation sodium is the predominant solute in the extracellular fluid (ECF). The concentration of sodium in the ECF is tenfold higher than inside all cells. The sodium cation together with its associated anions ("comp Anions") are the main determinants of ECF osmolarity. This electrolyte profile explains why total body sodium (sodium balance) determines the volume of the body fluid compartment of ECF.

The homeostatic control of ECF volume is thus tightly linked to the regulation of sodium balance. The student should give attention to the different set of homeostatic factors that govern ECF sodium *concentration* which are not identical to effectors that control ECF sodium *balance*.

Since the ECF and blood volume are determinants of blood pressure, mechanisms that involve renal functions in ECF volume and sodium balance are part of the overall, integrated physiology of blood pressure regulation. The kidneys excrete urine containing varying sodium and water load. The urinary excretion of sodium is a controlled event and tied to blood volume/pressure maintenance. Just remember this urinary effector of the kidneys' role in blood pressure (BP) as *BPee!*

The heart protects itself from being overloaded with fluid, i.e., overcongestion when blood volume expands. The cardiac chambers have mechano-volume receptors that monitor the "fullness" of the systemic vascular compartment. This is related to maintaining an optima central venous or right atrial pressure for normal circulatory venous return. The cardiac muscle also secrete a natriuretic hormone that acts on the kidneys to produce natriuresis, which is increased urinary sodium and water excretion.

Sodium is thus "so-dium" important in the reno-cardiovascular homeostasis of ECF volume which affects blood volume and pressure.

1. What common membrane protein reabsorbs sodium at the proximal tubule and the loop of Henle?

Answer The sodium/potassium ATPase is present at both the basolateral membrane of the proximal epithelial and loop endothelial cells, and this ATPase actively drives the transepithelial sodium reabsorption.

H. M. Cheng, *Physiology Question-Based Learning,* DOI 10.1007/978-3-319-12790-3_15

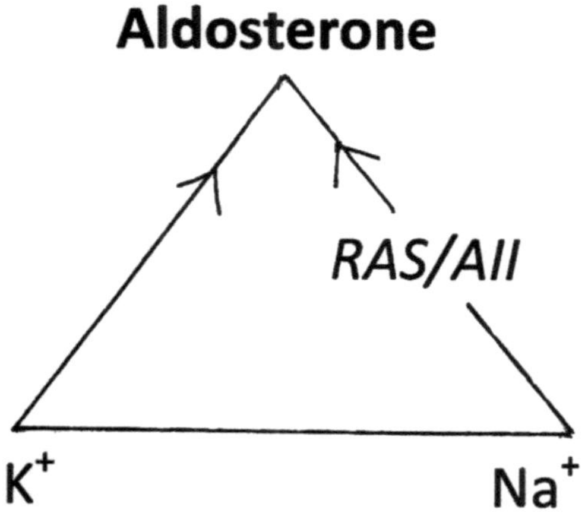

The adrenal gland (*triangle*!)secretes mineralocorticoids, steroid hormones, the chief species being aldosterone. Aldosterone is the key hormonal regulator for both potassium and sodium homeostasis. The action of aldosterone is for both K^+ and Na^+ balance, at the target principal cells of the collecting duct. Changes in ECF K^+ is directly sensed by the adrenal cortex while sodium imbalance is detected via a change in blood volume/pressure

Concept Filtered sodium is on average 99 % recovered or recycled into the peritubular circulation. The sodium is predominantly reabsorbed transepithelially along the nephron. Some sodium can diffuse through the tight intercellular junctions from the lumen to the interstitium. The entry of tubular fluid sodium into the cells uses different transporters at the luminal membrane. From inside the cells, the extrusion of sodium across the basolateral membrane into the renal interstitium employs the same ubiquitous Na/K ATPase pump.

At the proximal tubule, sodium enters the epithelial cells mainly through coupled transport with solute. This includes the familiar sodium-glucose cotransporter, amino acid-sodium symporter. A major luminal membrane exchanger, the Na/H antiporter also brings sodium into the cell.

At the thick ascending loop of Henle, the luminal sodium transporter is coupled to two other ions—chloride and potassium. We can correctly view this transporter as a component of the sodium-linked, secondary active transport of chloride and potassium.

At the distal tubule, we have a sodium-chloride cotransporter, and this membrane protein is the target site for the thiazide diuretics.

The sodium reabsorption at the principal cells of the collecting ducts is under hormonal regulation by the steroid, aldosterone. Here, sodium enters the principal cell via sodium channels at the luminal membrane. These membrane sodium channels are upregulated by aldosterone action. The basolateral membrane Na/K ATPase then completes the second step of transcellular sodium reabsorption.

The luminal sodium channels at the principal cells can be inhibited by drugs like amiloride and triamterene. These agents that obstruct the sodium channels are

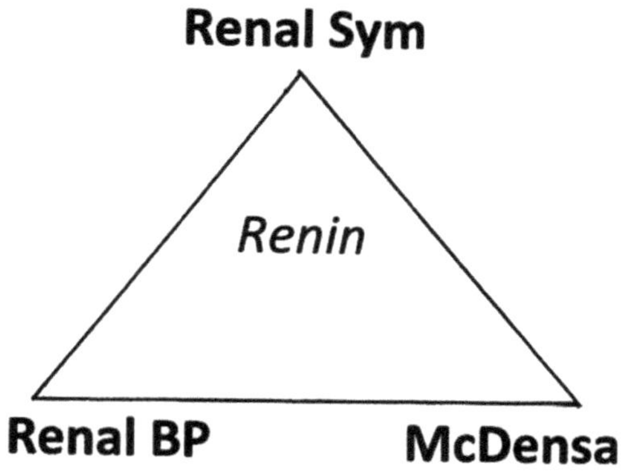

Renin. Three strokes to write the letter R. Three "stimulus strokes" affect renin secretion from the juxtaglomerular cells of the afferent arteriole: renal sympathetic nerve, renal arterial/perfusion pressure, and paracrine signal from macula densa

potassium-sparing diuretics, since they act on the principal cell (same cell that secreted potassium) and not "upstream" like the loop diuretics which has kaliuric side effects.

2. How are the movements for sodium and water related, if in any way, along the juxtamedullary nephrons?

Answer Due to the unique differential permeability to sodium and water along the juxtamedullary nephrons, water follows sodium reabsorption only at the proximal tubule.

Concept Water moves only when there is an available osmotic gradient. Along the juxtamedullary (jm) nephrons, the osmotic gradients are set up in different ways. At the proximal tubule, water is reabsorbed iso-osmotically following sodium/solute reabsorption. The latter generates the local osmotic transepithelial gradient for water to move transcellularly via aquaporins or paracellularly.

The renal medullary interstitium is uniquely hyperosmotic due to the renal countercurrent mechanism involving the jm nephrons and the associated peritubular vasa recta capillary. At the descending loop of Henle (LoH), the membrane is not permeable to sodium. Water alone is reabsorbed osmotically, driven by the hypertonic interstitium surrounding the descending loops.

At the ascending LoH, the membrane is unusual in being permanently impermeable to water. Sodium is, however, reabsorbed at the ascending LoH, passively at the thin segment and actively at the thick segment by the Na/K/2Cl cotransporter. The water remaining in lumen becomes progressively diluted and hypotonic, as it ascends the LoH.

The collecting ducts (also for the cortical nephrons) have changing permeability to water depending on the presence of the circulating hormone vasopressin which exits the vasa recta to bind to receptors on the basolateral membrane of the principal cells. Vasopressin stimulates the insertion of more aquaporins at the luminal membrane. Here again, the reabsorption of water is driven by the hyperosmotic interstitium as it occurs at the descending LoH.

Thus, in summary, if we imagine a cartoon of water molecule talking to sodium cation, the speech will be; at the proximal tubule, "I, water will follow you, Sodium." At the descending LoH, "You, Sodium can't follow me." At the ascending LoH, "I can't follow you, Sodium."

3. Do the renal tubules secrete sodium?

Answer There is no transepithelial secretion of sodium by the nephrons.

Concept The renal handling of sodium is excretion = filtration minus reabsorption. There is no transcellular active secretion of sodium by the epithelial cells along the nephron. In positive sodium balance, the kidneys compensate by decreasing sodium reabsorption. This leads to an increased sodium excretion.

What about passive secretion of sodium via the paracellular route? Quite certainly in the cortical nephrons, this should not be an event. This is, because there is no concentration gradient between the cortical interstitium and the tubular fluid.

At the proximal tubule (both cortical and jm nephrons), the surrounding tissue is the cortex which is iso-osmotic to plasma. The sodium reabsorption here is actually a net effect of transepithelial sodium movement and some back leak of sodium from the interstitial fluid into the lumen via the tight junctions. Normally, this back flux of sodium is minimum as water is reabsorbed iso-osmotically at the early proximal segment. Thus, the tubular fluid is still iso-osmotic to the interstitium. In glucosuria of diabetes mellitus, an osmotic diuresis results. Not only is water excretion increased as polyuria, there is also loss of sodium. This is due to the effect of unreabsorbed glucose in reducing the net sodium reabsorption. The back leak of sodium becomes significant, since water reabsorption at the proximal tubule is decreased due to the osmoactive glucose in glucosuria. The sodium concentration in the tubular fluid is then diluted and a sodium concentration gradient for passive sodium flux into the tubular fluid exist.

In the juxtamedullary nephrons, the LoH and the collecting ducts are surrounded by the osmotically stratified hypertonic renal medulla. At the ascending LoH, sodium is reabsorbed, and this sodium movement contributes to the generation of the hypertonic interstitium. At the descending LoH, we can expect some passive diffusion of sodium from the hyperosmotic interstitium paracellularly into the tubular fluid.

4. In hypertonic contraction, what parameter(s) activate the homeostatic mechanism for sodium balance control?

Answer The decreased volume stimulate sodium conservating mechanisms in order to restore the fluid volume lost.

Concept When you lose sweat, you are in both negative water and negative sodium balance. Sweat is a hypotonic fluid, but sodium is still lost in the perspiration. ECF volume is determined by the total body sodium (sodium balance).

In order to recover fluid volume, the loss of sodium has to be replaced. The kidneys will excrete less sodium. The mechanisms linking hypovolemia to decreased urinary sodium excretion involve the body sensing the volume reduction.

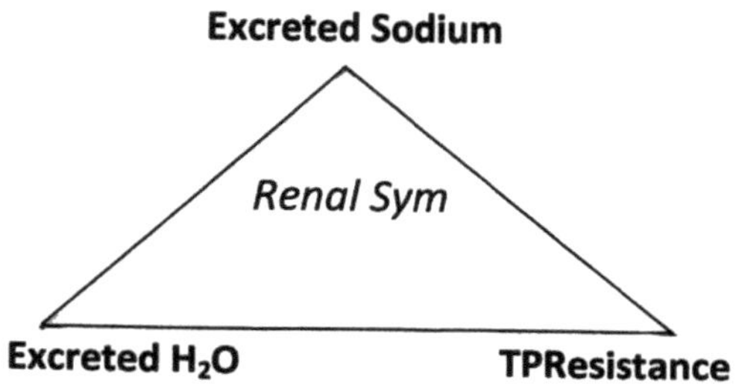

Renal sympathetic nerve (RSN) action conserves ECF/blood volume. This is effected as the RSN vasoconstriction decreases glomerular filtration rate (GFR). The filtered sodium and water are reduced. The RSN also increases renal tubular sodium reabsorption directly and also via stimulation of renin release. The RSN is also a part of the arterial blood pressure control effectors via renal arteriolar constriction, which raises the TPR

Three sets of volume sensors are operative—the high pressure arterial baroreceptors (carotid, aortic sinus), the volume receptors at the venous side, including the atria wall and pulmonary vasculature. In the kidneys the intrarenal baroreceptors sense the renal arterial pressure, which will be decreased with hypovolemia.

Note that the hypernatremia from sweating is not the parameter that will signal for sodium conservation. If it is, then the body should reduce sodium retention in order to respond to the hypernatremia. The negative sodium balance occurs as a result of lost body fluid, and all fluids contain sodium to varying extent.

The sensing of hypernatremia actually is by the hypothalamic osmoreceptors, since the sodium concentration is the main determinant of ECF osmolarity. Thus, the antidiuretic hormone (ADH) secretion will increase to help in restoring the water lost. But, the complete recovery to normal ECF volume will require the return of the total body sodium to normal.

Excretion of sodium is the filtered load minus reabsorbed sodium.

When the volume sensors detect hypovolemia/hypotension, the compensatory reduction in excreted sodium will be effected by decreasing filtered sodium and at the same time, increasing renal sodium reabsorption. The two major effectors that link volume sensing to urinary excretion of sodium are a nerve and a family of hormones. They are the renal sympathetic nerve and the renin-angiotensin-aldosterone "so-dium" conservative party (or *phy*arty in *phy*siology)!

5. In high sodium diet, what hormone action help to restore sodium balance?

Answer Natriuretic hormones are called into play to promote urinary excretion of sodium.

Concept A high sodium diet will lead to an expansion of the ECF volume. This is mediated by osmoreceptor/ADH mechanism and the thirst sensation. Because the osmoregulation is a rapid-responding feedback, the final equilibrium observed with a high sodium diet is an increased ECF volume rather than an increased ECF osmolarity.

Ten Commandments of OsMo(ses)

written on 2 *Tubulets*

1. Thou must note the different units for osmolarity (mOsm/L), osmotic pressure (mmHg) and tonicity (no units!).
2. Thou must know that sodium Cation concentration (with its associated 'compAnions') is the main determinant of ECF osmolarity.
3. Thou must not forget that osmotic pressure or 'effective osmolarity' is determined by the Non-membrane penetrating nature of the solute.
4. Thou should see for yourself that non-penetrating sodium ions, which also has the highest concentration in ECF are 'osmoactive' at the cell membrane but not at the capillary that separate the plasma from the interstitial fluid.
5. Thou should understand that at the capillary, plasma proteins that do not penetrate the capillary wall are 'osmoactive' and the protein concentration provides the oncotic or colloid osmotic pressure.
6. Thou should observe that changes in ECF sodium concentration (and thus ECF osmolarity) are almost always due to changes in Water balance and not to changes in total body sodium.
7. Thou should thus note that Control of ECF sodium concentration is also Osmo-regulation which is also related to control of Water Balance in the body.
8. Thou shall understand that Osmo-regulation involves the sensing hypothalamic Osmoreceptor /ADH secretion mechanism and the hypothalamic thirst center neurons.
9. Thou should see that ADH secretion and thirst sensation are affected by both ECF/blood osmolarity changes and also by ECF/ blood Volume changes (detected by volume/baro-receptors).
10. Thou shall thank God for your kidneys because the ADH collaborates with your kidneys in Osmo-regulation or Control of Water Balance by stimulating water conservation during negative water balance.

Osmotic force is the universal driving pressure for water movement throughout the body and certainly in the kidneys. Whenever an osmotic gradient is present or generated (different mechanisms operating at various sites), water will naturally move from the region of lower effective osmolarity (osmotic pressure) across a selectively permeable physiologic barrier (e.g., membrane, capillary wall) to the higher osmotic mmHg

With a high sodium diet, an initial hypertonic expansion (intracellular fluid (ICF)/ECF shift before ADH acts) will lead eventually to an isotonic expansion (after osmoregulation with ADH). This sequential explanation is helpful, but the student should note that the ICF to ECF fluid shift and ADH action overlap and occur dynamically.

An enlarged ECF volume will be detected by the various volume/pressure sensors. A general decrease in sympathetic activity including renal sympathetic input to the kidneys is a major compensation. Sodium excretion is enhanced.

In addition, volume sensors in the heart detect the increased vascular fullness and release atrial natriuretic peptide (ANP). This ANP is now part of a family of hormones including brain natriuretic peptide and the renal paracrine, urodilatin that, as the name suggest acts to increase both sodium and water excretion.

Circulating natriuretic hormones thus have opposing actions to aldosterone and ADH. In fact, the ANP has been shown to inhibit both aldosterone and ADH secretion. It likely also inhibits the renin release from afferent arteriolar juxta-glomerular (JG) cells in addition to the effect of reduced sympathetic stimulation on JG cells.

It can be reasoned that natiruretic hormones, including the paracrine urodilatin will increase the filtered load of sodium in order to increase sodium excretion. This is achieved as natriuretic hormones vasodilate the renal arterioles. The renal blood flow and hence the GFR is higher, producing the greater filtered sodium load.

6. How does renal sympathetic direct, nonhormonal actions on renal structures affect sodium excretion?

Answer The renal sympathetic nerve acts on the proximal tubule to increase the sodium reabsorption and on vasoconstricting the renal arterioles to decrease filtered sodium load.

Concept The more familiar action of renal sympathetic nerve (RSN) on renal handling of sodium to students perhaps is the effect on the arteriolar granular JG cells to secrete renin.

RSN conserves sodium. This autonomic nerve is sympathetic to sodium!

In other words, RSN decreases urinary sodium excretion. This is mediated by both reducing the filtered sodium load plus increasing renal sodium reabsorption. The renal arterioles are both constricted by the RSN. The renal blood/plasma flow is decreased, and GFR and hence filtered sodium load are lowered.

Concurrently, the proximal tubular epithelial cells are innervated by fibers of sympathetic nerve. Release of sympathetic neurotransmitter onto the proximal tubules stimulate sodium reabsorption.

The student may recall that when the intrinsic glomerulo-tubular (g-t) balance was described, the mechanism does not include the involvement of any extrinsic nerve or hormone action. In g-t balance, increase/decrease in GFR is compensated by a corresponding increase/decrease in proximal tubular reabsorption of sodium and water.

However, when the renal sympathetic nerve activity is high, the g-t balanced is overrided and pushed to the physiologic background. The RSN reduces GFR, but increases the proximal tubular reabsorptive function.

This occurs during negative sodium balance or hypovolemia and the physiologic priority to restore sodium/volume now is given over to the actions of RSN over g-t balance.

7. How is the macula densa involved in sodium balance?

Answer The macula densa has a paracrine signal that increases renin release from the afferent arteriolar JG cells.

Concept In the series of renal lectures, the macula densa (McD) is described as part of the juxtaglomerular apparatus (JGA) in effecting renal autoregulation of renal blood flow/GFR. In renal autoregulation, the paracrine from the McD is a vasoconstrictor that alters the vascular resistance of the preglomerular afferent arteriole.

The McD has a second role in renal functions and this is in the regulation of sodium balance. The second paracrine mediator from the McD modulates the secretion of renin from the JG cells. This McD paracrine renin stimulator(s) would be active when the person is in negative sodium balance of hypovolemia/hypotension. Renin then activates the generation of angiotensin II (AII), a potent vasoconstrictor and the trigger for aldosterone release from the adrenal glands.

Note that in the same situation, e.g., the hypovolemia which will reduce renal blood flow, the expected autoregulatory response should be vasodilation of the afferent arteriole. Thus, by physio-logic deduction, the renin is not involved in renal autoregulation since the AII that is derived from renin activationwill constrict the renal arterioles.

One could infer that the second sodium balance homeostatic signal of the McD is a more dominant paracrine than the paracrine that effects autoregulation.

The McD signal that releases renin from JG cell always works cooperatively and in concert with the other two inputs that affect JG cell renin secretion. These are the renal sympathetic nerve and the intrarenal baroreceptors (these volume/pressure sensors might be a structural component of the JG cells or a peri-JG cell receptor).

8. After drinking isotonic saline, how is sodium and water excretion affected?

Answer With the isotonic expansion, the sodium excretion is increased gradually, accompanied by water excretion. There is no peaked natriuresis.

Concept Some students in renal laboratory practical enlist a volunteer to drink several hundred ml of isotonic saline (how much depends on weight of student) in a short time. Then urine samples are collected every 30 min and measured for water and sodium excretion.

The subject will be in both positive sodium and positive water balance.

There is no observable water diuresis that is due to the inhibition of ADH secretion. The isotonic ECF will not affect ADH secretion, but the increased ECF/vascular volume will inhibit ADH. Since there is no observed evidence of a peak water excretion, inhibition of ADH cannot explain how the body will succeed in eliminating the excess water in urine. Imagine if the hypervolemia does initially inhibit ADH, any initial increased in water excretion will elevate the ECF osmolarity and this will begin to stimulate ADH and oppose the hypervolemic suppressing effect on ADH.

The positive sodium balance with the hypervolemia is dealt with by increasing sodium excretion. As sodium excretion increases over several hours, then water excretion will follow as osmoregulation also kicks in. The gradual compensatory response in excreting sodium reflects the time taken for reducing the plasma levels of the hormones in the renin-angiotensin-aldosterone family.

A decrease in renal sympathetic action due to hypervolemia-baroreflex feedback response also plays a part in promoting sodium urinary excretion.

9. How might renal atherosclerosis affect arterial blood pressure?

Answer The renal artery stenosis leads to hypersecretion of renin and hypertension

Concept The intrarenal baroreceptors respond to changes in renal arterial pressure. If they sense a reduction in the renal perfusion pressure, they will activate the JG cells to secrete renin. A reduction in vascular volume in blood volume loss will provoke the intrarenal baroreceptors to increase the renin secretion.

If the blood volume is normal, narrowing of the renal artery can also decrease the downstream renal perfusion pressure. This renal stenosis can be due to atherosclerotic vascular tissue growth. The resultant increase in renin will produce increased plasma angiotesin II and aldosterone.

Sodium and water retention will occur due to the combined renal actions of both aldosterone and AII. The hypervolemia will give a greater cardiac output, a determinant of arterial blood pressure.

AII is a strong vasoconstrictor. The total peripheral resistance, the other determinant of blood pressure, will be raised and this elevates the arterial blood pressure.

The unregulated secretion of renin in renal arterial stenosis will continue as long as the renal baroreceptors keep sensing the low renal perfusion pressure. The expanded ECF/blood volume will decrease the renal sympathetic stimulation of the JG cells. However, JG cells will still be "poked" by the low pressure at the afferent arteriole and also by positive signals from the macula densa. The decrease in renal blood flow due to the stenosis will lower the GFR and the volume of tubular fluid arriving at the distal tubular macula densa.

Secretion of natriuretic hormones as a result of the vascular congestion may provide some "aldosterone escape" effect to counteract the retention of sodium and water.

10. How is sodium balance altered in primary hyperaldosteronism?

Answer The positive sodium balance and expansion of the ECF volume inhibits renin secretion from the kidneys.

Concept In normal persons, the activation of hormonal stimulation of sodium conservation stars with renin. The "once upon a time" begins with renin release. In the pathophysiologic situation of a hypersecreting adrenal tumour of the glomerulosa cells (Conn's syndrome), sodium retention occurs in an unregulated manner.

ECF volume and blood volume will expand. The increase in cardiac output elevates the arterial blood pressure. The three (can the students recall from the previous questions?!) inputs that impinge on the JG cells to stimulate renin secretion are halted. Plasma renin concentration is reduced. So in Conn's syndrome, the "once upon a time" begins with high aldosterone secretion.

This causes the feedback inhibition of renin secretion from the JG cells.

The excess plasma aldosterone would also increase the loss of potassium in the urine with increased potassium secretion. Hypokalemia is an associated problem.

In the opposite endocrine condition of primary adrenal insufficiency (Addison's disease; deficiency or "minus" all adrenal corticosteroid hormones in this "Add" disease!), the mineralocorticoid aldosterone is severely lacking. The ECF will eventually be hypotonically contracted. This ECF status is not possible in a normal person with functioning osmoregualtion. Loss of sodium in the absence of aldosterone leads to a marked reduction in blood volume. This initial isotonic contraction will progress to a hypotonic contraction, when the hypovolemia becomes critical.

At this stage, the priority in maintaining a sufficient blood volume to sustain blood perfusion becomes more essential than osmoregulation. The threshold point for water reabsorption by the osmoreceptor/ADH mechanism is reset to a lower plasma osmolarity. This osmostat change is to allow more water to be reabsorbed before the ADH secretion is inhibited again.

Part IV
Cardio-Respi-Renal Physiology

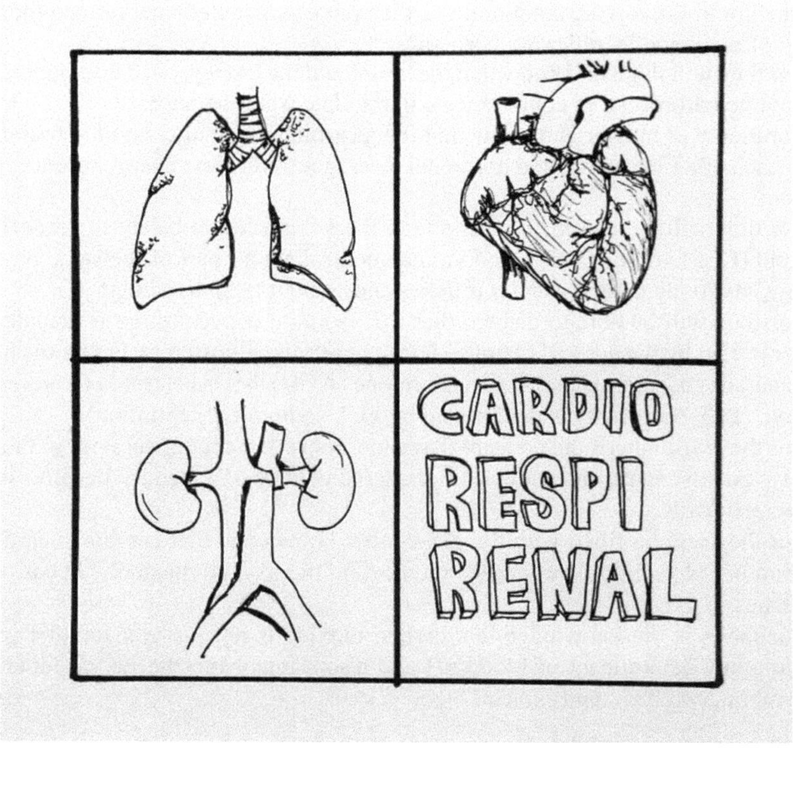

Blessed are the Integrated, a Physiologic Sermon on the Mount

A big picture, bird's eye view, of whole body physiology is very helpful to appreciate the interconnected, and concerted functions of the multi-physiology systems. Let me give some examples of integrated understanding by putting it in a fun way, modifying an account in the Gospels. This would be the metaphysiological narrative which will tell the story of our homeostatically designed human body. You might call this the "gospel-truth" of physiology.

Blessed are the integrated

1. For they will see that the pH of the extracellular fluid is protected by chemical buffer systems, respiratory, and renal functions.
2. For they will wonder at the cardio–renal overlapping mechanisms of blood pressure control involving autonomic cardiac/renal sympathetic nerves and the family of renin and its offspring hormones.
3. For they will thirst and know that the brain[1] and the kidneys, in combination with volume/baroreceptors communicate to regulate water balance.
4. For they will hunger and know that the gastrointestinal tract and the brain[2] talk to each other via neural and hormonal messages to regulate energy balance in the body.
5. For they will be able to show and tell others that sodium balance, extracellular fluid (ECF) volume, and blood volume control are all part of the same physiology, involving cardiovascular reflexes and sodium renal handling.
6. For they will be able to discern that ECF sodium concentration is actually not controlled by the adrenal gland sodium-conserving aldosterone but by the hypothalamus vasopressin antidiuretic hormone (ADH) that is central to Osmoregulation. (ECF osmolarity is determined by ECF sodium concentration).
7. For they will inherit the concept that water recycling occurs not only in the kidneys but also in the intestines from the large volume of aqueous, digestive juices secreted daily.
8. For they will be filled with the cell-centric knowledge that combined cardiorespiratory physiology are focused on meeting the oxygen/metabolic needs of the tissues.
9. For theirs is the knowledge that carbon dioxide is not just a metabolic waste, but a key determinant of blood pH and also a major metabolite vasodilator of cerebral, coronary, and skeletal blood vessels.

[1] Thirst center in hypothalamus.

[2] Orexigenic/anorexigenic hypothalamic neurons.

Chapter 16
Cardiorespiratory Physiology

What did Cardio say to Respi? "You provide the gas, I will provide the transport!" Cardiovascular physiology is appropriately taught before respiratory physiology in a well-designed undergraduate curriculum. We always teach the general before mentioning the exceptions. Pulmonary hemodynamics have their own unique functional characteristics, different from systemic control of arterial blood pressure and regional organ flow.

The cardiorespiratory integrated role is to provide well-oxygenated blood for tissues to match changing metabolic demands at rest or during physical activities. Metabolic carbon dioxide is also conveyed through blood to the lungs to be expelled. Blood CO_2, however, is not merely a metabolic waste. Blood/extracellular fluid pH is regulated through the essential function of alveolar ventilation/chemoreceptor sensing/CO_2 mechanisms.

1. How does the hemodynamic relationship, Flow $= P/R$ in the pulmonary circulation compare with the systemic circulation?

Answer The pulmonary flow rate is the same as the left ventricular output. The right ventricular moderate systolic pressure is matched by a lower pulmonary vascular resistance.

Concept The systemic and pulmonary circulations are in a circuit in a closed cardiovascular system. The left ventricular systolic pressure has a high maximum of 120 mmHg to propel blood into the periphery. The right ventricle is less muscular and need only generate an intraventricular pressure of ~30 mmHg to provide the right ventricular cardiac output.

This leads to the obvious derived fact that the pulmonary vascular resistance (pulmonary vascular resistance, pvr, is significantly less than the systemic "total peripherial resistance (TPR)."

There are also different cause and effect mechanisms between the pulmonary and systemic hemodynamics. In the systemic circuit, the blood pressure is determined by the two parameters cardiac output (CO) and TPR. Changes in CO and/or TPR are involved in the compensatory responses to maintain blood pressure. In other words, the relationship is a right to left direction.

H. M. Cheng, *Physiology Question-Based Learning*, DOI 10.1007/978-3-319-12790-3_16

The CO_2 C/O_2ckwise. Venous PCO_2 is the same as tissue PCO_2 (46 mmHg) as de-oxygenated blood has equilibrated with metabolic CO_2 in cells. At the lungs, the removal of CO_2 by diffusion and alveolar ventilation returns the PCO_2 in arterial blood to that of alveolar air (40 mmHg)

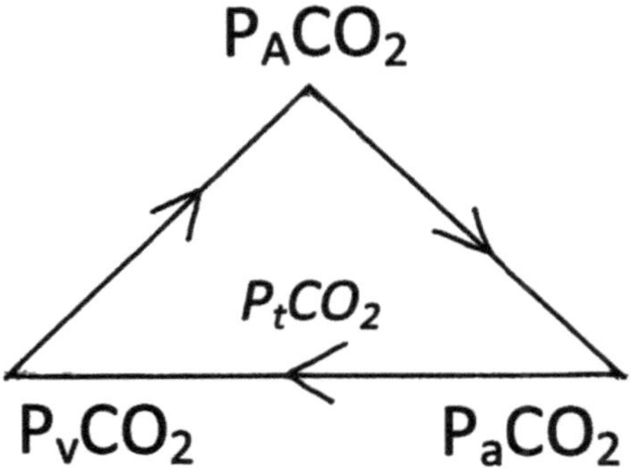

In contrast, in the pulmonary circulation, the pvr is mechanically affected by the pulmonary arterial pressure because the pulmonary vessels are relatively compliant. When the right ventricles eject a greater stroke volume, the raised pulmonary arterial pressure distends the pulmonary blood vessels. Besides vascular distention, the increased pulmonary blood pressure also recruits more underperfused, relatively nonpatent vessels at rest.

Thus, in pulmonary hemodynamics, the mechanistic pathway is from left to right Pulmonary arterial pressure = Right CO × pvr.

This unique feature of the pulmonary vascular explains why the pulmonary arterial pressure does not rise very much when the cardiac function increases during physical activity (note this reason is actually a right to left cause and effect).

2. Compare the rate of pacemaker action potential activity in heart and lungs and the muscle type stimulated.

Answer The cardiac atrial pacemaker cells emit action potentials at a rate of ~70 times/min. The respiratory "pacemaker" neurons generate action potential to the skeletal respiratory muscles at a rate ~10–12 times/min.

Concept The cardiac specialized cells at the right atrium spontaneously generate action potentials (AP) that are then transmitted to the rest of the myocardium. The inherent rate of AP production is modulated by inputs from both the parasympathetic and sympathetic nerves. The resting heart rate is dominated more by the vagal parasympathetic tone.

Normal respiratory rate is set by the spontaneous AP generating neurons in the brain stem. The impulses are conveyed to motor neurons that supply the inspiratory muscles of the lungs. Inspiration is an active muscle contraction event. Normal expiration begins with the cessation of the AP and the chest wall and elastic lungs recoil passively.

The skeletal muscles of inspiration are involved during voluntary hyperventilation. The initiating impulse originates from the cortical areas of the brain and it is transmittted directly to the respiratory skeletal muscles, bypassing the rhythmic regulatory neurons at the brain stem that sets the resting tidal volume and breathing frequency.

When one holds her breath, the inhibitory cortical impulses must be suppressing the autorhythmic signals from the brain stem to the respiratory muscles.

3. How does the driving perfusion pressure for pulmonary blood flow compare at the apex and base of the lung?

Answer The arterial-venous driving pressure at each level of the upright lung is the same since gravity affects both the hydrostatic pressure at both the arteries and veins.

Concept The gravity affects the distribution of the right heart cardiac output to the upright lung. The pulmonary blood flow is more to the base of the lungs and progressively decreases vertically to the apex.

This reduction in blood flow to the apex of the upright lungs is not due to any drop in the arterial-venous perfusion pressure. The hydrostatic pressure in both the arteries and the veins decreases proportionately above the level of the heart.

The resistance to pulmonary blood flow in the lungs is also modified by the intraalveolar pressure (P_A) that exerts a mechanical compression on the alveolar capillaries. The upright lungs is divided into three zones based on the essential driving pressure for blood flow that is operative at the different zones.

In the bottom third of the lungs (zone 3), the perfusing pressure is the arterial (Pa)-venous pressure (Pv) difference. Thus, here $Pa > Pv > P_A$.

In the middle third of the lungs (zone 2), the arterial pressure is greater than the alveolar air pressure which is now a pressure that exceeds the venous pressure

Here, $Pa > P_A > Pv$. The driving perfusing pressure is between the pulmonary arterial blood and the alveolar air pressure.

In the upper third of the lungs (zone 1), the alveolar air pressure can exceed the reduced arterial pressure and occlude the pulmonary blood flow The determining pressure difference that defines blood flow to the apex of the lungs is the $P_A > Pa$.

If arterial pressure is decreased due to hypovolemia or if the alveolar air pressure is increased as in positive pressure breathing, then the alveoli in Zone 1 will be ventilated by not perfused. These alveoli in these conditions no longer participate in gas exchange; they have a ventilation/perfusion (V/Q) ratio of infinity (Q = zero) and represent abnormal alveolar "dead" spaces.

4. Why does the pulmonary vascular resistance change during a respirator cycle? Alveolar, extra-alveolar vessels.

Answer The total pulmonary vascular resistance (pvr) is the sum of the vascular resistances at the alveolar and extraalveolar capillaries. The resistances of the two segments of pulmonary capillaries are differently affected by intrathoracic and intraalveolar air pressure changes during a respiratory cycle.

Concept Obviously the systemic total peripheral resistance (TPR) does not change during a respiratory cycle. The vascular radius of the pulmonary vessels, however, is affected by respiratory pressure changes during a breathing cycle.

The pulmonary capillaries that enmesh the alveoli are subject to the alveolar air pressure that opposes the intravascular pressure. Thus, with increasing lung volumes, the more expanded alveoli will compress and mechanically increase the vascular resistance.

In contrast, with inspiration, the intrathoracic pressure decreases. The radius of the extraalveolar blood vessels are affected by the transmural pressure which now involves the intrathoracic pressure outside the extraalveolar vessels. A deep inspiration will increase the transmural pressure and distend and lower the vascular resistance.

Thus, inspiration is associated with a relative higher resistance in the alveolar and a lower resistance in the extraalveolar pulmonary blood vessels. And in expiration, the resistance in the alveolar vessels becomes less and is now higher at the extraalveolar vessels.

The net change in vascular resistance during a respiratory cycle sums up to a lowest pvr at the end of normal expiration, at functional residual capacity (FRC). The pvr changes in a U-shape profile, where below and above FRC, the pvr increases with decreasing and increasing lung volumes, respectively.

5. In the upright lung, how does the partial pressure of oxygen and blood oxygenation at the apex compare with the base of the lung? V/Q high apex, but oxygenation more at base.

Answer Although the partial pressure of oxygen is higher at the apex of the upright lung, the net oxygenation is still better at the base of the lung, where pulmonary perfusion is more.

Concept The partial pressure of both oxygen (PO_2) and carbon dioxide (PCO_2) in the alveoli is not merely due to the alveolar ventilation but rather to the matched V/Q balance. In the upright lung, the V/Q matching varies from less than 0.8 (0.6) at the base of the lungs to more than 0.8 (3.0) at the apex of the lungs. The value 0.8 is the overall V/Q of the whole lungs with alveolar ventilation at 4 L/min and pulmonary blood flow at 5 L/min.

A higher V/Q at the apical alveoli means that they are relatively overventilated. As such, these alveoli will have a higher PO_2 (>100 mmHg, ~130 mmHg) and a lower PCO_2 (<40 mmHg, ~30 mmHg).

However, when we look at actual oxygenation of the pulmonary blood, the relatively underventilated alveoli at the base of the upright lungs are still better off. The alveolar PO_2 at the base of the lungs is about 90 mmHg.

The cooperative association of hemoglobin for oxygen means that at 90 mmHg, the blood oxygen saturation of hemoglobin will only be little affected. And, since

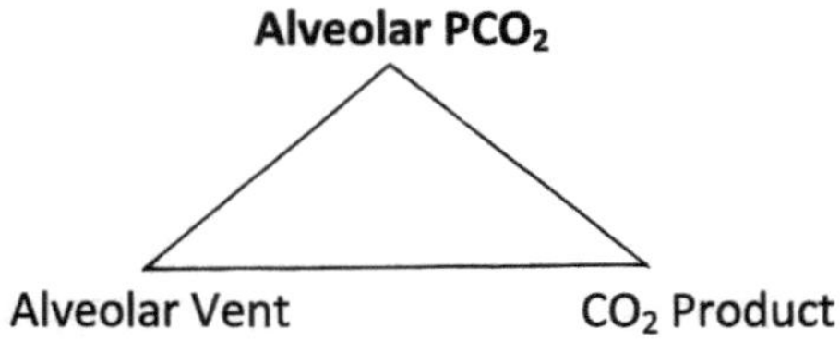

The partial pressure of alveolar air is determined by two events; rate of cellular CO_2 production (VCO_2) and the rate of alveolar ventilation. If VCO_2 is constant at rest, a voluntary hyperventilation will decrease the alveolar PCO_2. A respiratory alkalosis with hypocapnia is produced

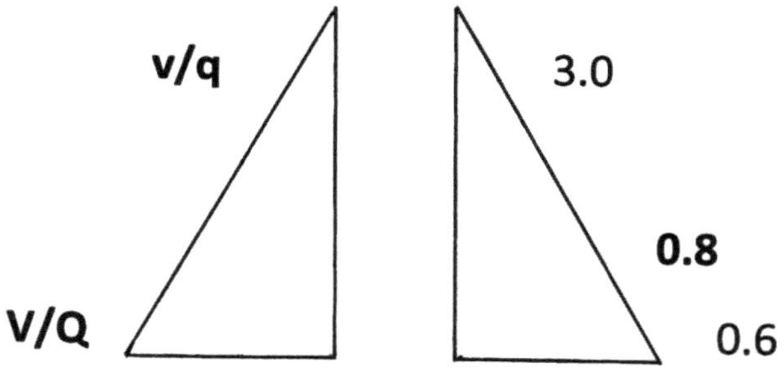

The alveolar ventilation (V) and the pulmonary blood perfusion (Q) decrease from the base to the apex of the upright lung. The overall V/Q of the whole lung is 4 or 5 L/min, giving 0.8. Since the reduction in Q is more pronounced than for V, the V/Q ratio increases from the basal to the apical zones

the oxygenation is perfusion limited, the better blood flow to the base of the lungs results in greater oxygenation.

The difference between basal and apical lung oxygenation is minimized during physical activity when the cardiac output is increased. The perfusion to the apex of the lungs is improved. The apical alveolar V/Q will be less than that at rest, but the oxygenation of the pulmonary capillary blood will be better.

6. At an intrapulmonary shunt, how will the alveolar air partial pressure of O_2 and CO_2 change? Alveolar dead space.

Answer At an alveolar shunt, the V/Q ratio is zero, meaning the alveolus is unventilated. The partial pressure of the respiratory gases will resemble that in the mixed venous blood, i.e., for oxygen 40 mmHg and for CO_2, 46 mmHg.

Concept There are a variety of shunts, the term used for both normal and abnormal cardiorespiratory physiology. In normal physiologic shunt, this refers to the slight mixing of shunted, deoxygenated blood with the main bulk of oxygenated blood from the alveoli. This normal admixture lowers marginally the partial pressure of O_2 in the arterial blood (<15 mmHg) from the blood that leaves the alveoli (~ 103 mmHg).

This physiologic shunt includes part of the bronchial blood and the blood in some of the small cardiac Thebesian veins that drain directly into the left ventricle, bypassing the lungs.

Thus, a physiologic shunt is due to anatomic architecture!. Clinically, when an "anatomical shunt" is named, it represents an excess of deoxygenated blood mixing with arterialized blood from the lungs. The classic example is a right to left shunt that is due to intracardiac septal defect. Not all the venous blood is delivered to the lungs to be reoxygenated. The portion that short-circuited from the right to the left atrium will dilute the PO_2 of blood that is pumped out by the left ventricle.

An intrapulmonary shunt, due to blockage of some airway passages, the affected alveoli are not ventilated (V=zero). The alveoli are not replenished with new, fresh inspired air. Eventually, the alveolar air at the end of the occluded airways

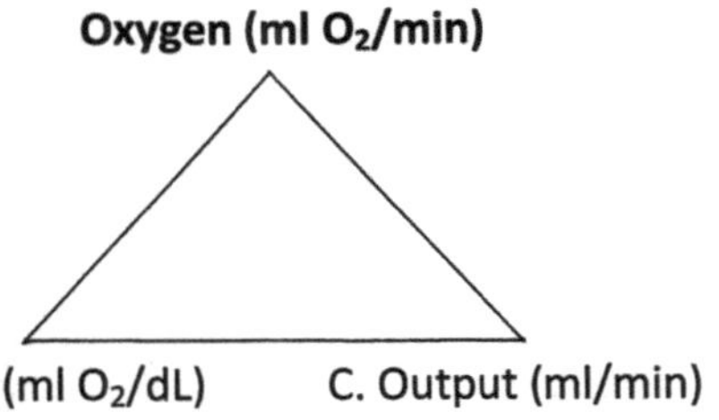

The rate of oxygen delivery to tissues is equal to the product of the arterial oxygen blood content and the cardiac output (CO). The CO can increase several folds during physical activity but the CaO_2 cannot even be doubled. Thus CO is the main determinant of increased O_2 delivery to the tissues during exercise

will equilibrate with the partial pressures of the gases in mixed venous blood PO_2, 40 mmHg and PCO_2, 46 mmHg.

Alveolar "dead space" in contrast occurs when there is a pulmonary embolus to a region of the lungs. The affected alveoli have very high V/Q ratio (infinity as perfusion is zero). Since no removal of oxygen from or release of CO_2 into the alveoli take place, the unperfused alveoli will only be ventilated and the alveolar air partial pressures of O_2/CO_2 will approach that of inspired air.

7. Does hyperventilation contribute mainly to the increase in oxygen requirement by cells during exercise? *Co x oxygen content sigmoid perfusion limited.*

Answer The increased pulmonary blood flow is the main contributor to the greater oxygen delivery rate to the cells. The hyperventilation, instead, has the essential role in the removal of metabolic carbon dioxide during exercise.

Concept Visibly, the person can be seen to have a deeper and higher breathing rate during exercise. The increased cardiac function is hidden from view. Quite naturally, since "seeing is believing," the student is apt to assume that the hyperventilation is the determining event that ensures the increased oxygen and energy supply to the tissues.

The lung oxygenation at rest is "perfusion limited." This term can be quite confusing and not quite descriptive of what it is meaning to say. This is not to say that the resting rate of oxygen diffusion across the alveolar-capillary membrane is inadequate or limited by blood flow.

"Perfusion limited" means that any further increase in lung oxygenation (at rest, 250 ml/min, the same value as tissue oxygenation or tissue extraction of O_2) can only be possible if the perfusion is not limited, "de-limited" or unlimited by increasing the pulmonary blood flow.

It is worth reminding that at rest, an equilibrium is achieved between PO_2 in alveolar air and pulmonary blood (103 mmHg) in half the alveolar capillary transit time. This is the reason for the perfusion-limited characteristic. Further, an additional transfer of "passenger" oxygen in the lungs is still possible and easily achieved, if more deoxygenated, mixed venous blood is "driven" to the "alveolar transit lounge" to pick up and transport more oxygen to the cells.

Then the question is "what is the point of 'having and catching more breaths' during physical activity?" The hyperventilation is not without a purpose. The essential role of increased alveolar ventilation is to maintain the carbon dioxide partial pressure in the blood. Metabolic CO_2 will make the blood acidic, producing a metabolic acidosis. The respiratory compensation via the chemoreceptor reflex is the control of the blood pH. The blood pH is reflected and determined by the concentration ratio of the components of the chemical buffer system. In the case of the bicarbonate/carbonic acid buffer

pH ~ bicarbonate/carbonic acid.

In the lactic acidosis of heavy exercise, the bicarbonate concentration is reduced. Compensatory hyperventilation will lower the PCO_2 to decrease the carbonic acid component. Blood pH is thus preserved close to physiologic 7.4.

The alveolar ventilation equation relates the arterial blood PCO_2 with the cellular CO_2 production (VCO_2) and the alveolar ventilation (V_A), where

Arterial $PCO_2 \sim VCO_2/V_A$

Thus, during moderate exercise, the arterial blood PCO_2 is relatively constant. The increased release of metabolic CO_2 is matched by the parallel increase in alveolar ventilation that expels the CO_2.

The observant and inquiring student will wonder and question "What then provides the stimulus for increased alveolar ventilation during exercise if the major chemical stimulus PCO_2 is unchanged?" This question should stimulate thinking and may cause you to hyperventilate!

8. In reduced hematocrit, how do the partial pressures of oxygen and CO_2 change, if any at the pulmonary capillary? Both unchanged oxygenation less, CaO_2 decreased Anemia….no stim of chemo…same as CO poisoning.

Answer Both the PO_2 and the PCO_2 of pulmonary capillary blood remain unchanged since the partial pressures are due to the dissolved fraction of the respiratory gases.

Concept The dissolved oxygen in plasma (PO_2) determines the percentage of oxygen saturation of hemoglobin. However, the reverse cause and effect is not true, i.e., the amount of oxygen bound to the hemoglobin does not affect the PO_2.

Thus, in anemia when the hematocrit is low, the decreased total oxygen that is bound by hemoglobin does not alter the amount dissolved in solution in plasma.

Since the PO_2 in low hematocirt anemia is unchanged, there will be no observed aortic/carotid chemoreceptor stimulation of increased breathing in the patient.

The total oxygen content is, of course, less when the hematocrit is decreased. The total amount of O_2 that diffuses down its partial pressure gradient in the lungs is higher. But, the diffusion PO_2 gradient is unchanged across the alveolar-capillary membrane (103 and 40 mmHg at entry point, respectively).

In carbon monoxide toxicity, the CO competes with its >200 times higher affinity for hemoglobin than as for Hb–O_2. The blood oxygen content is drastically reduced by CO competition that wins hands down (*or hbO$_2$ down*!). The dissolved oxygen in plasma, however, is untouched by CO. Therefore, the chemoreceptors do not sense any decreased PO_2 and breathing is not labored or there is no sign of air hunger.

The person who has CO poisoning will slowly "peter" away and lose consciousness. This is a common choice method to take one's life as it can be unfortunately ended quite "comfortably."

Would the reduced hematocrit affect CO_2 since red cells are essential for converting metabolic CO_2 from the tissues to plasma bicarbonate? This conversion is catalyzed by the red cell enzyme, carbonic anhydrase (C@). Since this is a catalytic event, the decrease in hematocrit should not cause any measurable reduction in the conversion of CO_2 to bicarbonate. The mixed venous blood partial pressure of CO_2 should still be 46 mmHg. In the different case when the C@ enzyme is inhibited, the venous PCO_2 can rise.

9. Why is the right heart failure an occurrence in long-term ascend to high altitude?

Answer Right ventricular hypertrophy develops due to the pulmonary hypertension that is caused by general hypoxic pulmonary vasoconstriction.

Concept The afterload for the left ventricle is the mean aortic or arterial pressure. The afterload for the right ventricle is the pulmonary arterial pressure. The afterload represents the pressure against which the ventricles have to pump or generate sufficient pressure for ejection of stroke volume.

Hypoxic pulmonary vasoconstriction (hpv) is a unique vascular design that serves the function in optimizing V/Q matching. Should there be a local regional mismatch of V/Q where the alveoli are underventilated, the decreased alveolar PO_2 will produce the hpv so that the pulmonary blood flow is not "wasted" on the underventilated alveoli but channeled to well-ventilated alveoli.

This adaptation to regional lung hypoxia at sea level presents a problem when the person is exposed for a long time to lower atmospheric oxygen at high altitudes. The hypoxia is now a general condition throughout the lungs. Thus, we can expect that the hpv is no longer restricted to a certain area but widespread at every alveolus.

The general increased pulmonary vasoconstriction will raise the pulmonary arterial pressure.

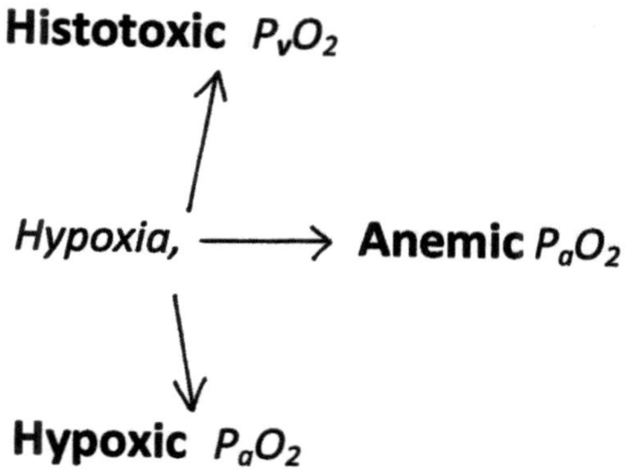

The cells can be deprived of sufficient oxygen (hypoxia) in various ways. In hypoxic hypoxia, the partial pressure of O_2 (PO_2) in blood is reduced. In anemic hypoxia, the hemoglobin-bound and thus the total O_2 content is decreased with no change in the PO_2. If the cells are metabolically suppressed, the venous blood PO_2 is characteristically elevated

The right ventricle will, with time, hypertrophy, in response to the increased cardiac work against the elevated afterload. This early compensation of the cardiac myocardium may not be adequately sustained in the longer period. The right ventricle may begin to weaken under the constant need to generate a greater ventricular tension to maintain a normal cardiac output to the lungs.

The right ventricle can eventually decompensate with resulting pump failure.

10. What compensatory changes in red blood cells (RBC) help to improve oxygen delivery to the tissues in low atmospheric environment? Number, **2,3 dpg....extreme altitude alkalosis incr affinity**.

Answer The glycolytic metabolite, 2, 3 diphosphoglycerate (DPG) in the red cells increases and the hemoglobin–oxygen affinity is reduced to promote more unloading of oxygen to the cells.

Concept The cardiorespiratory functions are directed at the final cell consumer. The physiology is "cell-centered" (I suppose "self-centered!"). Adaptations in physiology is aimed at ensuring that the functioning unit of physical life, the cell is nourished adequately for health and physical activity. The neuron and the cardiomyocyte are the two most essential cell types that are the focus of physiological, homeostatic attention.

Under hypoxic conditions, the intracellular red cell DPG concentration increases. DPG decreases the affinity of hemoglobin for oxygen. The main effect is to increase more unloading of oxygen to the cells.

It might be thought that the upload of oxygen in the lungs will be less with the reduced $Hb–O_2$ affinity. However, because of the unique sigmoid, cooperative $Hb–O_2$ interactions, even a drop of arterial PO_2 at 60 mmHg will only reduce the $Hb–O_2$ saturation, just a few percentage to 90 %.

At extreme altitudes, when the atmospheric PO_2 becomes critically low for survival, an increased $Hb–O_2$ affinity effect might be a needed compensation. This is caused by the respiratory alkalosis from hyperventilation at the high place.

Alkalosis shifts the oxy–hemoglobin curve to the left with increased $Hb–O_2$ affinity. DPG has the opposite action in a rightward shift of the $Hb–O_2$ curve. At extreme heights, the alkalotic effect to increase uptake of oxygen in the lungs takes precedence when the alveolar air PO_2 falls too much.

Chapter 17
Cardiorenal Physiology

The heart and the kidneys are intimate partners in handling pressure—arterial blood pressure. The cardiac muscle and renal nephrons both receive sympathetic nerve inputs which serve as part of the integrative, wired connections between them. The kidneys also secrete hormones, renin, and erythropoietin which are "wireless" hormonal signals that maintain the blood volume. Arterial baroreceptors that monitor blood pressure/volume are found in the carotid artery/aorta and the preglomerular renal arteriole. There are homeostatic linkages between sodium balance, blood volume, blood pressure, and urinary excretion of sodium. Regulated renal tubular secretion of potassium and excretion in the urine ("pee") is essential to maintain normal cardiac pacemaker activity. The three Ps here (pee, potassium, pacemaker) are bound together for life.

1. Which determinant of arterial blood pressure equation is affected by renal handling of sodium?

Answer The cardiac output determinant of blood pressure is dependent on blood volume which in turn changes with the sodium balance (total body sodium).

Concept The two determinants of blood pressure are total peripheral resistance (TPR) and cardiac output. Total body sodium is the main determinant of extracellular fluid (ECF) volume/blood volume. Blood volume changes will affect the mean systemic filling pressure and hence the venous return.

Sodium balance changes alter the blood volume because sodium is the main determinant of ECF osmolarity. For example, if a person goes on a high sodium diet, the ECF osmolarity soon changes. Osmoregulation is a rapid physiological response mechanism involving the hypothalamic osmoreceptors/antidiuretic hormone (ADH) and the hypothalamic thirst neurons.

The normalization of ECF osmolarity due to the positive sodium balance leads to new equilibrium of an increased ECF/blood volume.

Eventually the body will activate compensatory mechanisms for the hypervolemia/positive sodium balance to restore blood volume.

2. How do the Starling's forces at the peritubular capillary serve its function?

Answer The balance of Starling's forces at the peritubular capillary favors reabsorption along the length of the peritubular capillary.

H. M. Cheng, *Physiology Question-Based Learning*, DOI 10.1007/978-3-319-12790-3_17

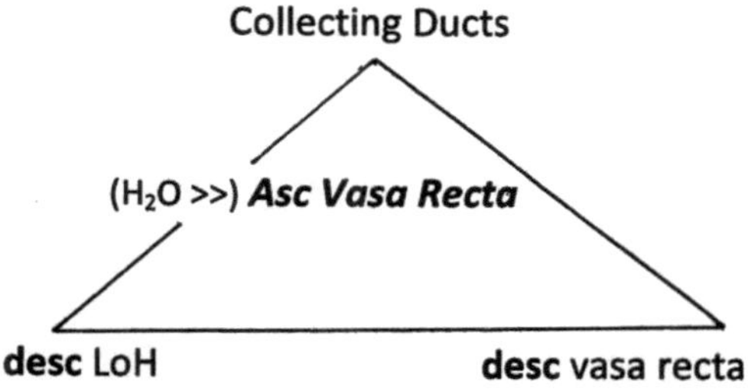

In the renal medulla, water that is reabsorbed from the descending loop of Henle (LoH), the descending peritubular capillary, vasa recta, and the collecting ducts (under ADH action) are all returned to the circulation via the ascending vasa recta. There is no dilution or disturbance to the hyperosmotically stratified medullary interstitium

Concept Renal handling includes filtration, reabsorption and secretion. The latter two tubular processes involve the peritubular capillary. Reabsorption of water and solutes is the movement across the epithelial cell, transcellularly and/or paracellularly from the tubular fluid in the lumen into the peritubular capillary.

The peritubular hydrostatic pressure is relatively low due to the high vascular resistance of the postglomerular efferent arteriole. The glomerular hydrostatic pressure is high at ~ 50 mmHg and drops to less than 20+ mmHg in the peritubular capillary.

In contrast, the plasma oncotic pressure in the peritubular capillary is high at ~40 mmHg. The end-glomerular blood has an elevated oncotic pressure (>35 mmHg) due to high filtration fraction (20 %). The blood that exits the efferent arteriole enters the peritubular capillary. Therefore, the peritubular capillary plasma oncotic pressure is greater than its hydrostatic pressure. The net Starling's force will be a reabsorptive force.

This is a favorable physiologic situation as a large portion of filtered solute is reabsorbed at the proximal tubule. An equally high fraction of filtered water (~75 %) is then reabsorbed following the solute reabsorption by isoosmotic water reabsorption.

3. How does a baroreflex affect renal handling of sodium?

Answer A baroreflex induced sympathetic discharge will increase renal sympathetic akction to effect decreased sodium urinary excretion.

Concept Increases in sympathetic nerve action tend to increase cardiac function and vascular resistance. These combine to raise arterial blood pressure.

Hypotension or hypovolemia will trigger the baroreflex to increase general sympathetic discharge. Among this increased effector neural action is the renal sympathetic nerve. Physiologically renal sympathetic action will, in concert also, help to raise the blood pressure.

The positive effect on blood pressure by renal sympathetic activity acts via the conservation of sodium. Sodium-dependent blood volume mechanism provides the link to the regulation of blood pressure.

Water reabsorption at the kidneys and the intestines are driven by the osmotic gradient. The osmotic pressure gradient is generated by prior solute transepithelial movement from the lumen to the interstitium. Water is reabsorbed also at the intestines as the large fluid volume in all the gastro-intestinal (GI) secretions (saliva, stomach, pancreas) has to be recovered. Water moves both transcellularly and paracellularly at the kidneys and the gut

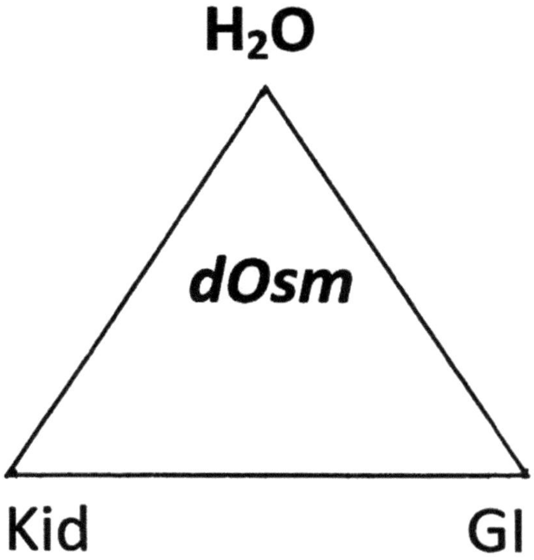

There are a variety of pathways that renal sympathetic action conserves sodium. One is by reducing the filtered sodium load since the glomerular filtration rate (GFR) is decreases by sympathetic constriction of the renal arterioles. The sympathetic nerve fibers also innervate the proximal tubular cells and they have a direct effect in increasing sodium reabsorption.

Thirdly, the juxtaglomerular (JG) cells that secrete the hormone/enzyme renin is stimulated by sympathetic input. Activation of the renin-angiotensin system increased the plasma concentration of aldosterone, the key hormone that increases sodium reabsorption at the collecting ducts in the kidneys.

The intrarenal baroreceptors at the afferent arteriole also participate in compensation during decreases renal blood perfusion when the blood pressure decreases. However this is not via a neural baroreflex involving the brain stem. Intrarenal baroreceptors sense directly a reduced renal arterial blood pressure, and this leads to the increased release of renin from JG cells of the afferent arteriole into the blood.

4. What vascular action of renal sympathetic contributes to blood pressure control? Tpr

Answer The vasoconstriction of the renal arterioles contribute to the TPR. TPR is a determinant of arterial blood pressure.

Concept The word "total" in TPR is misleading. It does not imply all regional resistances to blood flow is changes, either increased or decreased at the same time. When the sympathetic discharge is increased during a baroreflex, the overall (better word than "total") peripheral resistance is increased. This occurs due to vasoconstriction by the sympathetic nerve to arterioles. However, this vasoconstricting effect is *selective* and the increased vascular resistance is tailored towards organs like the kidneys, gastrointestinal tract and the skin.

It will be noted that these organs are nonessential when we consider the critical importance of the cerebral and coronary circulations. Increased sympathetic action arising from baroreceptor activation does not raise the cerebral and coronary vascular resistance, for obvious reasons if the body is designed to keep us alive!

The renal sympathetic nerve vasoconstricts both the afferent and efferent arterioles. This is part of the "selective" vasoconstriction throughout the body to increase the "overall" peripheral resistance.

The segmented approach in teaching physiology is partly responsible for why our students commonly do not include the renal sympathetic nerve in TPR changes. The baroreflex is detailed during the cardiovascular system and frequently, the sympathetic efferent compensatory actions are described as actions on cardiac function, venoconstriction/venous return and "total" peripheral resistance (PR).

Seldom at this point is the renal arterioles stated and when renal arterioles are mentioned a few weeks later in the curriculum, the focus is generally on the renal arteriolar resistances and glomerular filtration.

Our teaching of physiology should be rightly "total" and integrated and not "selective" and compartmentalized.

5. How is the flow autoregulation in coronary and renal circulation effected by different mechanisms?

Answer Blood flow autoregulation to the heart is effected by myogenic and metabolic mechanisms and in the kidneys are mediated by myogenic and the macula densa (McD) tubule-glomerular feedback.

Concept Certain organs in the body exhibit intrinsic abilities to autoregulate its blood flow and are able to maintain a relatively constant flow within a defined range of blood pressure fluctuations. Hemodynamically, if the perfusion pressure changes, there will be a parallel change in the blood flow if the vascular resistance is not altered.

For the heart and the kidneys (and also the brain), this autoregulation is distinct and independent of extrinsic nerve or hormonal actions.

In these three organs, the smooth muscle of their arterioles demonstrates a myogenic property of responding to mechanical stretch. An increased stretch by pressure will produce a myogenic contraction of the blood vessel. This means that when the perfusion pressure increases, there will be a corresponding increase in the myogenic contraction. Since the flow is equal to pressure/resistance, the blood flow is autoregulated to a constant value.

In the coronary circulation, a metabolic feedback could also contribute to the autoregulation. For example, an acute decrease in perfusion pressure will reduce blood flow initially and this will lead to accumulation of metabolites in the myocardium. Several vasodilator metabolites will then act on the coronary arterioles and decrease the vascular resistance. This metabolic action cause by the initial decrease in perfusing pressure will ensure a normal coronary blood flow.

In the kidneys, a major intrinsic mechanism that autoregulates renal blood flow (RBF) is tied to its physiologic purpose of autoregulating the GFR. The structural complex involved in renal autoregulation (RenAg) is the juxtaglomerular apparatus

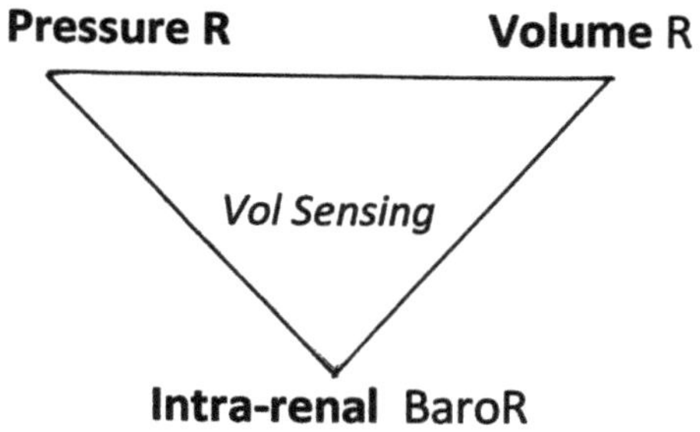

Blood volume sensing is related to monitoring ECF volume changes as the blood volume is a part of ECF. In the high-pressure arterial side, the carotid and aortic sinus pressure/baroreceptors monitor the blood volume/pressure. Volume receptors are found in the low-pressure part of the cardiovascular system at the great veins/right atrium and the pulmonary vasculature. The intrarenal baroreceptors at the preglomerular, afferent arteriole sense renal arterial pressure

(JGA). The JGA has two components, the preglomerular afferent arteriole and the distal tubular McD. In an indirect way, the McD senses changes in distal tubular fluid electrolyte when the RBF/GFR changes consequently on renal arterial pressure fluctuations. An initial increase in renal perfusion pressure will bring more electrolytes that are sensed by the McD (mainly chloride and sodium ions). The McD is excited and release a paracrine vasoconstrictor. This local vasoconstrictor diffuses to the neighborhood afferent arteriole and constricts the blood vessel. The increased renal perfusion pressure is thus balanced by a higher afferent vascular resistance. The RBF/GFR is normalized. This intrinsic RenAg mechanism (independent of the renal sympathetic nerve or circulating hormones) is a feedback from the distal tubular McD to the preglomerular afferent arteriole, hence, the name "tubular-glomerular feedback" or simply the McD sensing mechanism.

6. State with brief reasons if RenAG serves primarily a metabolic function?

Answer The maintenance of RBF and GFR are primarily for handling changes in ECF volume and electrolyte balance, in particular sodium and potassium. A decrease in RBF/GFR also has significant effects on secretion and excretion of metabolic products.

Concept The essential organs, heart and the brain are critically dependent on an adequate blood supply for its functions. The RBF obviously also provide energy source, nutrients and oxygen to the renal tissues. In contrast to the myocardium and the brain tissues, a major role of RBF (resting 20 % of cardiac output) is to ensure a normal GFR. A large portion of the oxygen consumption of the kidneys is in the active reabsorption of sodium and sodium linked solutes.

Fluctuations in GFR will result in parallel fluctuations in filtered load of the solutes and water. Filtered load is calculated by the product of GFR and the filtered solute concentration (which is the same as plasma concentration if the solute is freely filtered). An acute increase of GFR will present a higher load of tubular fluid to the distal nephrons which may exceed the membrane transport capabilities further downstream along the nephron.

The importance of an optimal tubular fluid delivery to the distal nephrons is highlighted by another intrinsic phenomenon besides the McD autoregulatory mechanism. RenAG of RBF/GFR is not perfect and some fluctuations do occur. A second event involving the proximal tubule will compensate for changes in the filtered load that enters the Bowman's capsule. This is called the glomerulo-tubular (G-t) balance (sounds like the tubular-glomerular feedback, but this G-t balance takes place at the proximal tubule.

An increased filtered load of solutes and water will be compensated by an increase in proximal tubular reabsorption. Similarly, a reduced filtered load will be accompanied by a decreased reabsorption of fluid and solutes at the proximal tubule. The G-t mechanism involves changes to peritubular capillary fluid dynamics involving changes in pericapillary Starling's forces. Changes in GFR will affect the hydrostatic and oncotic pressures in the peritubular capillary "downstream" from the glomerular capillary which is in series with it, connected by the efferent arteriole.

7. How does a poor cardiac output lead to increased sodium retention by the kidneys?

Answer Decreased perfusion of the kidneys is sensed, and this triggers an increased conservation of sodium in an effort to normalize "perceived hypovolemia" that causes the poor RBF.

Concept The kidneys are hard wired to ensure that the heart pumps a normal cardiac output. The communication "bloody" line between the heart and the kidneys is the blood pressure. A reduced cardiac output is insufficient to sustain an arterial blood pressure that is needed to circulate blood to the peripheral organs.

The kidneys detect this hypotension at the intrarenal baroreceptors at the afferent renal arteriole. The kidneys "perceive" any reduced perfusion pressure as possibly resulting from hypovolemia. As such, the renal mechanisms activated during poor cardiac output (although blood volume is normal) act to increase sodium and water retention.

Sensing of hypotension by the renal baroreceptors will release renin. The sequential progression of the renin-angiotensin pathway produce the bioactive angiotensin II. Angiotensin II is a primary stimulus of aldosterone secretion from the adrenal cortex. The angiotensin II (AII) and aldosterone both act to increase renal tubular sodium reabsorption.

AII also directly promote sodium reabsorption at the proximal tubule. Sodium retention is also contributed by a reduced load of sodium filtered. This is due to vasoconstriction of renal arterioles by AII. The AII vasoconstricts the renal arterioles and besides lowering sodium filtered load, this renal vasoconstriction also sums up with other regional resistances into the TPR.

A reduced cardiac output will activate the carotid/aortic baroreceptors. The effector sympathetic discharge from a baroreflex has identical actions as AII in decreasing filtered sodium load, increase TPR and stimulate proximal tubular sodium reabsorption. Renal sympathetic nerve action also stimulates renin secretion which has the end effect of decreasing urinary sodium excretion.

8. When you drink a large volume of water in a few minutes, does your kidney regulate blood osmolarity, blood volume or both? Give your rationale.

Answer Consuming a large volume of water leads to positive water balance. This is rapidly adjusted by an increased excretion of hypotonic urine via inhibition of osmoreceptor/ADH mechanism.

Concept Should the ECF volume change, the student must ask whether there is also any change in the sodium balance (total body sodium). This is because the control of sodium balance (related to ECF volume) by the renin-angiotensin mechanism (RAS), renal sympathetic nerve action does not need to participate physiologically if the sodium balance is unchanged.

This is the scenario in the hypotonic expansion from drinking water. Although the volume of the ECF has increased, it will be incorrect to state, e.g., that a compensatory inhibition of the RAS by the increases ECF/blood volume will be part of the compensation. This is merely a positive water balance with no change in the sodium balance. It is often mistaken when the parameters "sodium concentration" and sodium balance are used interchangeably. If RAS inhibition is involved, then an increased sodium excretion in the urine will follow. This will not make physiologic sense as the person has a normal sodium balance before drinking the water. If RAS inhibition occurs, the person will end up with a negative sodium balance although he drinks only water!

The body can rapidly respond to the positive water balance by osmoregulation. Osmoregulation of the ECF is rightly linked to controlling the sodium concentration of the ECF. This is because sodium concentration is the main determinant of the ECF osmolarity. Drinking an excess volume of water will lower the sodium concentration/osmolarity of the ECF. This produce inhibition of the hypothalamic osmoreceptors and the ADH/vasopressin secretion from the posterior pituitary is inhibited.

The urinary excretion of water increases in positive water balance. The urine is dilute and large in volume. Note that the urinary sodium excretion (amount/time) is unchanged, although the sodium concentration in the hypotonic urine is lower.

9. How does the renal control of electrolyte balance maintain normal cardiac electrical function?

Answer Perhaps the most important electrolyte regulated by the kidneys which affect cardiac activity is potassium.

Concept Renal function is essential for electrolyte balance in the ECF and this includes sodium, calcium and potassium (also magnesium which we know less of its established physiology). Sodium balance and volume of ECF/blood are interrelated and govern by renal mechanisms operating in concert with cardiovascular control of blood pressure.

Calcium ion is involved in the depolarization of the sinoatrial pacemaker cells and also in the prolonged depolarization phase of the ventricles. Hypoclacemia and hypercalcemia do have effects on cardiac function. The kidneys in combination

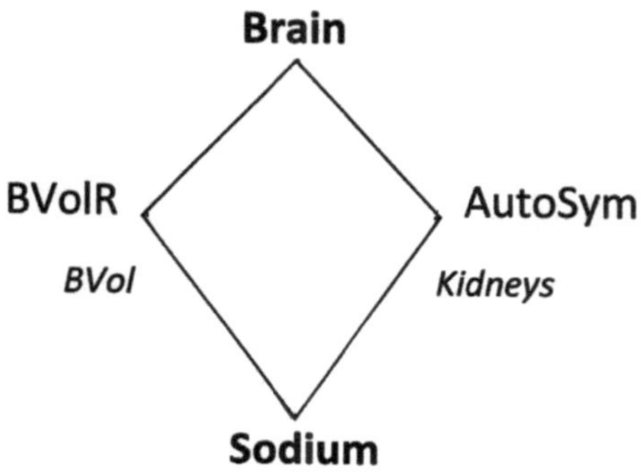

The brain and sodium balance. Blood volume changes are monitored, and sensor information processed in the brain stem. The autonomic sympathetic nerve activity is a major effector in the blood volume/pressure control. Increased/decreased renal sympathetic actions restore normal sodium balance, which is the main determinant of ECF/blood volume

with the triad of hormones, parathyroid, vitamin D and calcitonin participate in the homeostasis of calcium.

Clinically, changes in ECF potassium have profound effects on the myocardium. The resting membrane potential of a cell is due to the transmembrane potassium concentration gradient. In hypokalemia, cardiac cells are hyperpolarized and less excitable.

In hyperkalemia, a constant elevated ECF potassium may initially produce increased excitability of the myocardium. However, the hyperkalemia-induced partial depolarization will eventually lead to inactivation of the voltage-gated sodium channels. Critically, the heart can cease to beat.

ECF potassium, at a low regulated level of 4–5 mmol/L is monitored by cells that secrete aldosterone from the adrenal cortex. Hyperkalemia stimulates aldosterone release (postprandial hyperkalemia stimulates insulin also). Aldosterone acts on the principal cells at the collecting ducts to promote tubular secretion of potassium. Both the basolateral K/Na ATPase and the luminal membrane potassium channels are increased by aldosterone to enhance potassium secretion.

The homeostasis of potassium includes this hormonal fine-tuning of tubular secretion because on a normal daily diet excess potassium is added to the ECF.

10. How does proteinuria change the capillary fluid dynamics in skeletal muscle?

Answer Proteinuria decreases the plasma oncotic pressure and this tends to increase the net filtration at the capillary predisposing to edema.

Concept Traditionally, the explanation for peripheral edema due to proteinuria is described as follows: Loss of protein in the urine reduced the plasma oncotic pressure that is the key reabsorptive osmotic force at the capillary. The balance of Starling's forces favors a greater filtration at the arteriolar end of the capillary. At the venular end of the capillary, the recovery or reabsorption of fluid from the

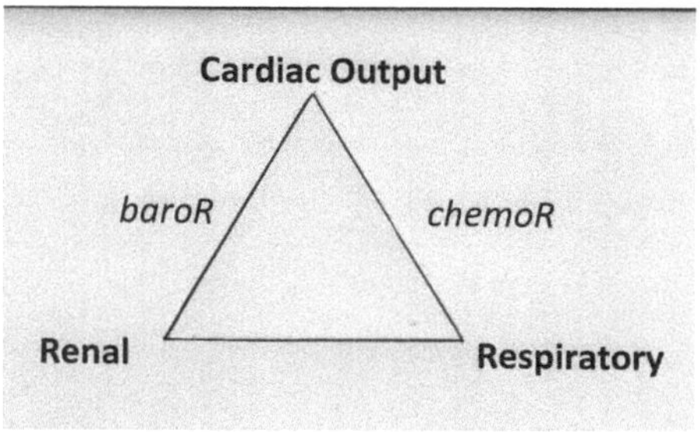

The cardiac output is monitored by the kidneys and respiratory function. The peripheral chemoreceptors (carotid/aortic bodies) can be stimulated in stagnant hypoxia. Decreased effective circulatory volume is sensed by the intrarenal baroreceptors at the afferent arteriole. This is the physiologic "Reno-Respi CO Sandwich"

interstitial space into the capillary is also decreased. Fluid, thus, accumulates in the interstitial compartment of the ECF.

The corresponding contraction of the vascular blood compartment is also viewed as a stimulus that then activates the volume-conservating renin-angiotensin system. This accounts for the sodium and fluid retention in the person.

Recently, alternative proposals for the link between proteinuria and development of peripheral edema have been raised.

Chapter 18
Respi-Renal Physiology

The lungs take in air and let out air. The kidneys take in fluid and let out fluid. The oxygenated blood from the lungs is circulated to the kidneys. The carbon dioxide in the arterial blood has a physio-logic role in modifying the renal tubular handling of bicarbonate and hydrogen ions. The lungs and the kidneys are major players in the daily military physio-drama of pH defence. The combined respiratory and renal strategies victoriously prevent any pH disturbance and maintain a stable optimal pH of the extracellular fluid (ECF).

Besides critical roles in pH regulation, the kidneys and the lungs release hormones that participate centrally in the control of arterial blood pressure. The cardiovascular system joins them in this physiologic triumvirate to ensure a constant perfusion pressure from the heart to all cells in the body.

1. Do the lungs or the kidneys excrete most of the daily acid load?

Answer The lungs eliminate most of the daily acid production which is carbonic acid from metabolism of foods.

Concept The daily production of acid is about 15 moles. A large portion of this daily acid load is carbonic acid from the hydration of metabolic CO_2. Normal respiratory function is thus essential to maintain the arterial blood pH of 7.4. A pH of 7.4 is equivalent to a hydrogen ion concentration of 40 nmol/L (1 nmol = 10^{-9} mol).

For noncarbonic acids produced from metabolism, the kidneys are the only route for excretion. The daily renal excretion of acid is much less at about 70 mmole. One millimole is still 1 million times the concentration of 1 nmol! Therefore, the renal function is critical for preservation of normal body fluid pH.

The time line for respiratory control of pH is earlier than that for renal excretion of acid. However, although renal compensations for pH fluctuations takes a longer time, the renal handling of acid and recovery of the main base, bicarbonate ion effectively rectifies and restores the ECF pH in response to any acid–base challenges.

H. M. Cheng, *Physiology Question-Based Learning*, DOI 10.1007/978-3-319-12790-3_18

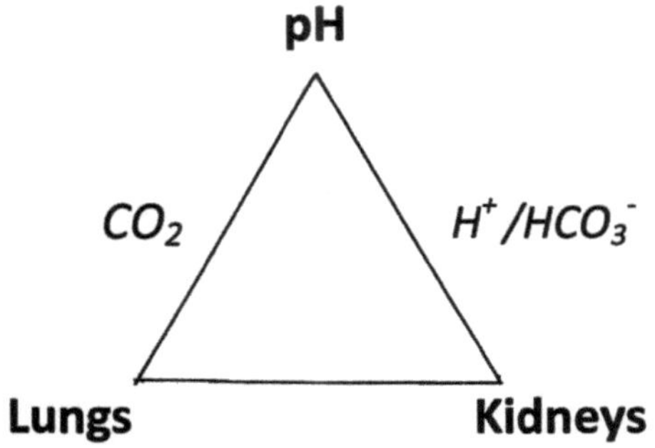

ECF/Blood pH is carefully regulated by chemical buffers that are integrated with respiratory and renal functions. The major daily acid load is carbonic acid and the lungs remove the excess metabolic CO_2. Noncarbonic acids are handled by the renal nephrons that secrete hydrogen ions and restore blood bicarbonate, the major base in the ECF

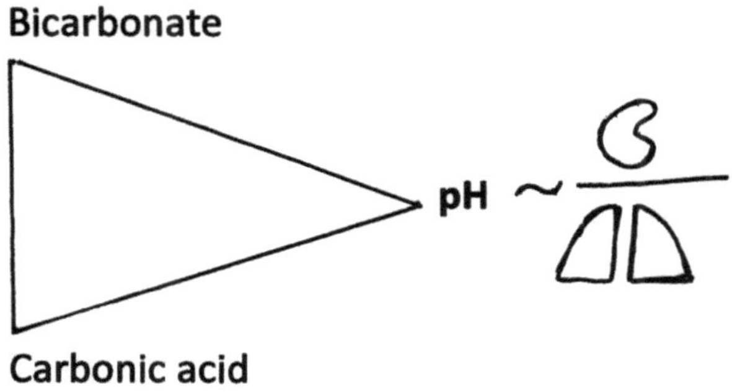

The pH of ECF is reflected and associated with the respective ratios of the base/acid pairs of all the chemical buffers. Quantitatively, the major ECF buffer is the bicarbonate/carbonic acid system. Renal function targets the bicarbonate component while respiratory function, via regulating PCO_2 affects the carbonic acid

2. How does a renal hormone compensate in two ways in pulmonary hypoventilatory conditions?

Answer The kidneys will secrete erythropoietin. The respiratory acidosis will also increase tubular acid secretion.

Concept In hypoventilation, the partial pressure of O_2 in arterial blood will decrease. The hypoxia will be detected by the kidneys and the renal hormone erythropoietin will be secreted. Erythropoietin will increase red cell production and a secondary polycythemia will develop.

This hypoxia-induced compensation in a higher hematocrit is also seen in during acclimatization to high altitude hypoxia. A large polycythemic compensation is counterproductive, however, as the increased hematocrit will increase the blood viscosity. Vascular resistance to blood flow will then be higher.

In addition, the kidneys will respond to the respiratory acidosis by increasing tubular hydrogen secretion. The secretion of more H^+ by the renal tubules is linked

The renal tubules secrete hydrogen ions. The tubular intracellular biochemistry that produces the protons is linked to the tubular reabsorption of filtered bicarbonate and also the synthesis of new bicarbonate to replace ECF bicarbonate ions that have been consumed during chemical buffering

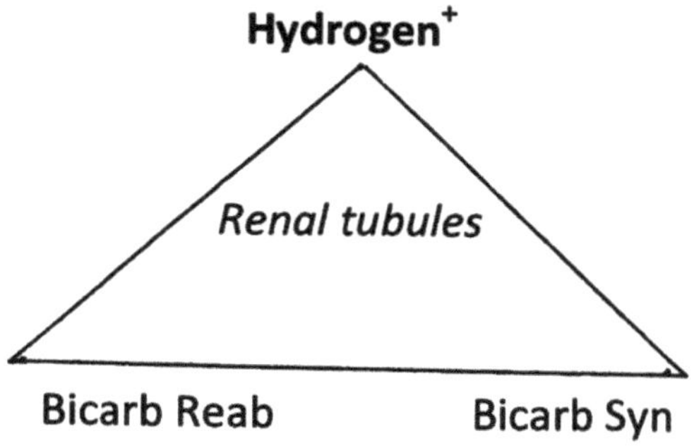

to replenishing the blood bicarbonate concentration. Both reabsorption and new synthesis of bicarbonate by the tubules is increased when the H^+ secretion is higher.

There is evidence that changes in ECF pH also affect some other hormones that act on the renal tubular secretion of H^+. Acidosis stimulates renal renin secretion. The generation of angiotensin II (AII) along the activated renin-angiotensin pathway helps to increase the blood pH. AII stimulates a hydrogen ATPase transporter in the collecting ducts. Hydrogen ions are also secreted when AII acts on the sodium hydrogen countertransporter at the proximal tubule.

Acidosis also stimulates the release of the adrenal steroid hormone aldosterone. Aldosterone stimulates a hydrogen ATPase to increase acid secretion by the nephrons.

3. How does arterial blood PCO_2 affect the renal tubular acid secretion?

Answer Increased arterial PCO_2 will increase tubular secretion of hydrogen ions in the kidneys.

Concept The renal tubular cells at the proximal tubule and at the collecting ducts secrete H^+.

The two main stimuli which affect tubular acid secretion is blood PCO_2 and blood pH. If PCO_2 is increased or pH is decreased, the renal tubules will secrete more H^+.

For CO_2, this is more easily visualized. CO_2 from the peritubular capillary enters the tubular cells at the basolateral side and is hydrated to carbonic acid by carbonic anhydrase (C@). The carbonic acid then dissociates into bicarbonate ions and H^+. The proton is secreted at the lumial membrane into the lumen.

For a low pH, the cell signaling mechanism is less obvious. We can envisage pH sensors on the basolateral membrane of the tubular epithelial cells. Detection of a reduced ECF pH by the basolateral border will then lead to increased cellular machinery to secrete more acid. These cellular effects of the low interstitial pH could include upregulating C@ activity and the appearance of more luminal sodium–H^+ exchange transporters.

The major ECF chemical buffer is the bicarbonate/PCO_2 buffer system. Quantitatively, this is the most efficient buffer as it is an "open" system. The "openness" of the system means that both the numerator bicarbonate and the denominator PCO_2

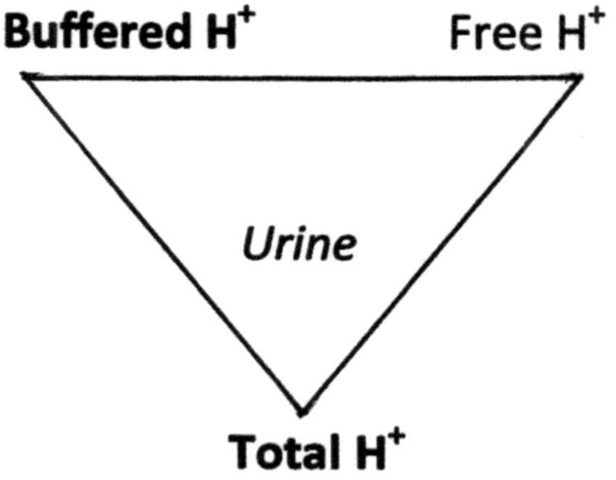

The total excreted hydrogen ions in urine is predominantly bound to urinary buffers, mainly phosphate and as ammonium ions. The measured urinary pH is due only to the free, unbound H^+. At the urinary pH of 4.4, this is still a small amount of H^+ (1000×40 nmol/L or 40 μmol/L of urine)

are linked to the renal function and pulmonary function respectively. This is sometimes written nicely as

$$\text{pH} \sim \text{Kidney/Lungs}.$$

A respiratory acid base disturbance will thus be compensated by renal tubular reactions. A respiratory acidosis during alveolar hypoventilation will lead to reduced renal secretion of acid. This also means that the tubular reabsorption and synthesis of HCO_3 is decreased. Similarly, if the acid–base fluctuation is due to a renal cause, respiratory adjustments will take place. If the pH disturbance is due to a nonrespiratory, nonrenal reason, then both the lungs and the kidneys can compensate, e.g., metabolic acidosis from diarrhea.

4. During vomiting, does the renal function compensate effectively?

Answer Vomiting leads to a metabolic alkalosis which is not effectively compensated by the kidneys due to contraction alkalosis.

Concept The loss of gastric acid during vomiting results in a metabolic alkalosis. The respiratory compensation will be a decreased alveolar ventilation due to a depressed stimulation of the arterial chemoreceptors.

The kidneys would be expected to reduce its tubular secretion of acid, produce an alkaline urine to compensate for the metabolic alkalosis. However, there is a paradoxical acidic urine in the metabolic alkalosis resulting from the loss of gastric acid.

This is due to the activation of the renin angiotensin system (RAS) by hypovolemia due to fluid loss in the vomitus. Two components of the RAS stimulate tubular secretion of H^+. This phenomenon of a hypovolemia-driven alkalosis is called "contraction alkalosis."

Fluid replacement in the sick person will restore the normal effectiveness of the kidneys to adjust to the alkalosis.

The increased blood pH will reduce the proton secretion, which is also accompanied by less tubular reabsorption and synthesis of bicarbonate.

Note that the respiratory reflex should increase the PCO_2. How then does this elevated PCO_2 not oppose the appropriate tubular response to metabolic alkalosis to secrete less acid? In metabolic alkalosis, the filtered load of bicarbonate is still high enough that an excess of bicarbonate still escapes reabsorption. In addition, if the alkalosis is severe, secretion of bicarbonate can also occur at specific intercalated cells at the collecting ducts.

5. Compare the functions of carbonic anhydrase in the respiratory and renal physiology.

Answer In red blood cells, the enzyme carbonic anhydrase is essential for the formation of bicarbonate, the major form of CO_2 transported in blood. In renal tubules, carbonic anhydrase is need for three processes, namely the reabsorption and synthesis of bicarbonate plus the secretion of hydrogen ions.

Concept The enzyme, carbonic anhydrase, C@ is the key catalyst in the hydration reaction of carbon dioxide. The C@ is minimum in the plasma and found in the cytoplasm of erythrocytes. CO_2 enters the red cells and is hydrated to carbonic acid, which dissociates to bicarbonate and proton. The bicarbonate exits the red cells in exchange with chloride while the H^+ is buffered by hemoglobin.

In renal tubules, the C@ is found both inside the cells and also on the cell membranes. Specifically, this is the enzymatic scenario at the proximal tubule. The C@ at the luminal membrane catalysed the reaction between secreted H^+ and filtered HCO_3. The reaction in the lumen proceeds to CO_2 and water. The CO_2 enters the proximal tubular cell is rehydrated by cytosolic C@. Inside the proximal epithelial cells, both bicarbonate and proton is generated. The proton is then secreted via the sodium/hydrogen antiporter into the tubular fluid. The HCO_3 also leaves the cell but at the basolateral membrane and is reabsorbed into the peritubular capillary. Note that the filtered bicarbonate is reabsorbed not directly at the luminal surface but by a "convoluted" indirect pathway as described.

The carbon dioxide can also enter cells at the basolateral side. This is the situation when we consider the tubular synthesis of *new* bicarbonate. Note that at the proximal tubular reaction above, the filtered bicarbonate is merely reabsorbed and there is no synthesis of new bicarbonate (the proximal ammonium synthesis does produce HCO_3). Also, the secreted H^+ at the proximal tubule does not become excreted in urine but is incorporated as water to be reabsorbed. The C@ catalysed pathway inside tubular cells of the collecting ducts is involved in both the tubular secretion of H^+ for the urinary excretion of ammonium and titratable acids, mainly urinary phosphate.

Interestingly, the gastric parietal cells that secrete hydrogen ions also uses the same C@. In parietal cells, CO_2 is hydrated and dissociated to produce H^+ and bicarbonate. The hydrogen is secreted into the gastric lumen as part of hydrochloric acid in gastric juice. The HCO_3, however, exits the parietal cell at the basolateral membrane. This accounts for the postprandial increase in blood pH also called "alkaline tide".

CO_2 is not merely a metabolic waste. CO_2 is the principal chemoregulator of normal breathing. Carbonic acid is a component of the major extracellular fluid bicarbonate/carbonic acid buffer. Tissue CO_2 is a key arteriolar vasodilator that effects cerebral and coronary blood flow autoregulation, and also regional tissue active and reactive hyperemia

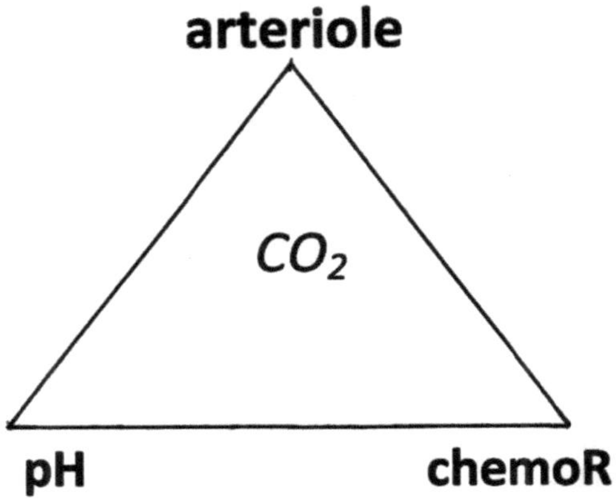

6. How do the kidneys alter urine composition to help to ascend mountains?

Answer The urine becomes alkaline with more excretion of bicarbonate. This resensitizes the central chemoreceptors that were inhibited by the high altitude, hypoxia-induced respiratory alkalosis.

Concept Carbon dioxide is the dominant chemical regulator for respiration. During an ascent to high altitude, hypoxia is the primary stimulus for increased ventilation. The increased breathing soon produces a respiratory alkalosis due to more CO_2 being removed. The hypoxic hyperventilatory response is thus opposed by the hypocapnia.

Acclimatization towards a better ventilatory response takes place over the next few days. The hypocapnia raise the pH of the interstitial fluid that surround the chemoreceptors in the brain stem.

The sensitivity of the chemoreceptors is improved when this pH is decreased. This is effected by urinary excretion of bicarbonate. The decreased partial pressure of CO_2 will cause the renal tubular cells to decrease bicarbonate reabsorption.

Mountain climbers can speed up the acclimatization process by taking carbonic anhydrase inhibitor. The reabsorption of filtered HCO_3 at the proximal tubule is dependent on carbonic anhydrase (C@), present at both the luminal membrane and inside the tubular cells. Inhibition of C@ will lead to an increased renal clearance of bicarbonate in the urine. This is renal compensation for respiratory alkalosis.

A minor expected effect would be a slight increase in acidity of the venous blood. This is because the red cell C@-catalysed production of bicarbonate will be suppressed.

7. Compare the effects of sympathetic activity on airflow and renal blood flow?

Answer The sympathetic nerve acts to produce bronchodilation and vasoconstrict the renal arterioles.

Concept The effect of autonomic sympathetic nerve action on bronchial smooth muscle is mainly indirect via stimulating the secretion of adrenal caecholamines, adrenaline/noradrenaline. Adrenaline acts on the same beta adrenergic receptors on airway smooth muscle that is bound by neurotrans mitter noradrenaline releases from postganglionic sympathetic fibers. An airway resistance is decreased with bronchodilation. This effect makes sense when we think of the general increase in sympathetic activity during exercise when breathing rate is increased.

The effect or renal sympathetic nerve on renal arteriolar smooth muscle is to produce vasoconstriction. Renal blood flow and hence glomerular filtration rate (GFR) is decreased. If we think of the exercise scenario again, the increased sympathetic arteriolar constriction has a role is redistributing the cardiac output to the skeletal muscles where metabolism is higher. The renal vasoconstriction is an alpha adrenergic receptor binding effect.

The renal arteriolar constriction also has a function in maintaining arterial blood pressure by affecting the "total" peripheral resistance (TPR). The blood vessels are dilated in the exercising skeletal muscles, and this reduces the TPR. The blood pressure is still maintained by a greater cardiac output and selective vasoconstriction in organs including the renal and splanchnic vasculature.

The mesangial cells at the glomerular region are smooth muscle-type cells and they respond to vasoactive agents. Mesangial cell contraction will decrease the total area available for filtration and lower the GFR.

Renal sympathetic action in the kidneys has a key role in conserving sodium. These effects on sodium balance in turn regulate ECF/blood volume and operate via a reduction in filtered load of sodium and stimulating secretion of renin and its antinatiruretic hormones in the renin-angiotensin system.

8. State the effect, if any of hypoxia on pulmonary and renal vasculature.

Answer In the lungs, hypoxia causes a unique pulmonary vasoconstriction. The renal tissues are relatively tolerant to hypoxia.

Concept In the lungs, the blood vessels do not respond in the usual way as seen in other tissues. In all other organs, when blood oxygen supply is inadequate to meet metabolic demand, the tissue hypoxia will compensate by vasodilating the blood vessels. Hypoxic pulmonary vasoconstriction (HPV) is thus not a response to ensure sufficient oxygen to the pulmonary alveoli (the lungs are filled with O_2-rich air; perhaps the bronchial vasculature that supplies nonalveoli structures has similar hypoxic-vasodilation response).

The physiologic pulmonary rationale for the HPV is to maintain an optimal ventilation/perfusion matching at the alveoli exchange area.

The renal blood flow has a major role in providing a normal, large GFR (180 L/day; note plasma volume is around only 3 L). The tissue oxygen extraction for the kidneys is mainly used to provide energy for the tubular transport processes. The key epithelial cell transporter at the basolateral membrane is the Na/K ATPase. Since a lot of solutes are transported at the tubules via sodium-linked mechanisms, there is a proportionate relationship between renal oxygen consumption and the rate of tubular sodium reabsorption.

Hypoxia, of course, increases the renal erythropoietin production and release.

9. How is pulmonary blood flow and renal blood flow determined by using the same experimental principle?

Answer Fick's principle is used for determining pulmonary blood flow, using oxygen as the measured parameter and for renal blood flow, the solute measured is p-amino hippuric acid (PAH).

Concept Fick's principle is used to estimate regional blood flow. The equation states that the organ blood flow is equal to the rate of "extraction" of the solute (*Es*) divided by the arterial-venous concentration difference (Sa—Sv) of the solute, i.e.,

$$\text{Flow} = \text{Es/Sa} - \text{Sv}.$$

For pulmonary blood flow, *Es* is the rate of oxygenation ml O_2/min, and the oxygen content difference between pulmonary venous/arterial blood is used.

For renal blood flow (RBF), the renal plasma flow is determined and the RBF calculated from knowing the hematocrit. In addition, the value obtained for renal plasma flow is underestimated by about 10 % since not all the RBF is delivered to the glomeruli for filtration.

RBF is estimated by using the renal clearance of PAH.

Es is the extracted (excreted load) of PAH. At a small concentration of PAH used, the venous PAH concentration is zero as PAH, an organic acid, will all be secreted by the renal tubules. Thus, the flow formula is simplified to excreted load of PAH/ arterial PAH concentration. This converts actually to the renal clearance of PAH.

10. Is respirator disturbance of acid–base balance associated with any pH associated flux of potassium across the renal tubular membranes?

Answer Respiratory causes of acid–base imbalance generally do not produce any transmembrane potassium shift since the primary disturbance is in CO_2.

Concept Carbon dioxide is lipid-soluble and moves freely across cell membranes. Therefore, there is no need for K^+ exchange across the membranes to preserve electroneutrality.

This noninvolvement of potassium shift is also seen in several forms of metabolic acidosis. In lactic acidosis, the lactate is available to enter the cell with the hydrogen ion, and electroneutrality is preserved.

If hydrogen ions enter the cells, and this is not accompanied by any anion, then the intracellular potassium cations efflux to maintain electroneutrality. Thus, in this situation, acidosis or academia produces a secondary hyperkalemia.

At the renal tubular epithelial cells, this efflux of K^+ into the peritubular capillary also means that the secretion of K^+ into the tubular fluid is less. The "efflux" may in reality be due to an inhibition of Na/K ATPase by acidosis which will reduce the pumping of K^+ into cells.

In metabolic alkalosis, the reverse cation exchange between H^+ and K^+ can take place. The intracellular buffers, organic phosphates and proteins can release H^+ which exits the cells. Potassium enters cell and hypokalemia is precipitated by the alkalosis.

Students who are discerning will note that for potassium to enter cells, it must move against a steep concentration gradient, at least 30 times higher inside the cells (ECF 4 mmol/L, intracellular fluid (ICF) 140 mmol/L). This indicates that the K/H transmembrane exchange is a more complex event and not merely a function of a definite membrane protein antiporter.

Bibliography

Bransford J, Brown A, Cocking R (1999) How people learn: brain, mind, experience and school. National Academy Press, USA
Cheng HM (2013) Thinking through physiology. Pearson, Malaysia
Cheng HM (2014) Conceptual learning in physiology. Pearson, Malaysia
Cheng HM, Durairajanayagam D (2012) Misconceptions highlighted among medical students in the annual international intermedical school physiology quiz. Adv Physiol Educ 36:229–232
Cheng HM, Mah KK (2014) Cardiovascular physiology–figure-based instructions. Cengage Learning
Costanzo L (2010) Physiology, 4th edn. Saunders, Elsevier
Deschenes G, Feraille E, Doucet A (2003) Mechanisms of oedema in nephrotic syndrome: old theories and new ideas. Nephrol Dial Transplant 18:454–456
Erickson L (2008) Stirring the head, heart and soul: redefining curriculum, instruction and concept -based learning, 3rd edn. Corwin Press
Hall J (2011) Guyton and Hall textbook of medical physiology, 12th edn. Saunders, Elsevier
Kibble JD, Halsey CR (2009) Medical physiology. The big picture. McGraw-Hill
Raff H, Levitzky M (2011) Medical physiology. A systems approach. McGraw-Hill
Sherwood L (2010) Human physiology. From cells to systems, 7th edn. Cengage Learning

H. M. Cheng, *Physiology Question-Based Learning*, DOI 10.1007/978-3-319-12790-3